*Give Us This Day* . . .

# GIVE US THIS DAY

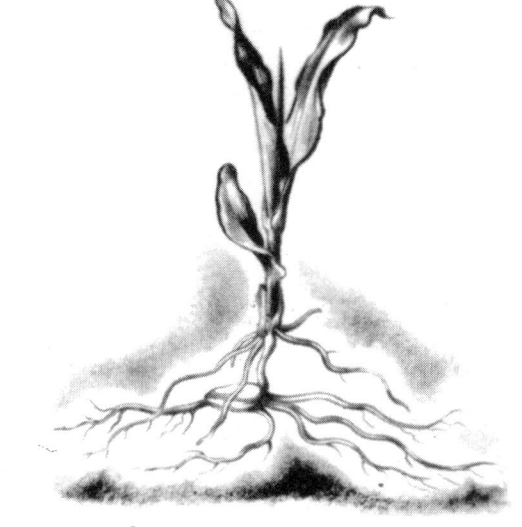

### BY Clare Leighton

*Essay Index Reprint Series*

 BOOKS FOR LIBRARIES PRESS
FREEPORT, NEW YORK

Copyright, 1943, 1971, by Clare Leighton

Reprinted 1971 by arrangement with
Harcourt Brace Jovanovich, Inc.

---

Library of Congress Cataloging in Publication Data

Leighton, Clare Veronica Hope, 1900-
   Give us this day.
   (Essay index reprint series)
   1. Country life--U. S.  2. Agriculture--U. S.
I. Title.
S521.L442  1971     917.3'03'917     76-152187
ISBN 0-8369-2513-0

---

PRINTED IN THE UNITED STATES OF AMERICA
BY
NEW WORLD BOOK MANUFACTURING CO., INC.
HALLANDALE, FLORIDA 33009

*To the "Powerful Valley" and the
Vale of Aylesbury; because
they are so much alike*

## Table of Contents

| | Page |
|---|---|
| PROLOGUE: THE POWER OF THE SEED | 1 |

*Chapter*

| | | |
|---|---|---|
| I. | THE ARTERIES OF SUPPLY | 5 |
| II. | THE FARMER | 13 |
| III. | CORN AND WHEAT | 23 |
| IV. | MEAT | 35 |
| V. | TRUCK FARMS | 44 |
| VI. | FRUIT | 52 |
| VII. | MILK AND EGGS | 59 |
| VIII. | FARMS | 67 |
| | EPILOGUE: THE EARTH REMAINETH | 76 |

# PROLOGUE:
# THE POWER OF THE SEED

The sun beats down upon the shipyard. It beats upon welder and riveter. It strikes the great sheets of steel, blinding the men's eyes with the dazzle of mirrored suns. It beats upon the battleship that soon will be launched.

The steel battleship dwarfs her builders, until they look like an army of ants crawling over a carcass. They work with the activity of ants. They work in the madhouse of a wartime shipyard, choked by the fumes of welding, deafened by the fusillade of the pneumatic riveters.

This giant battleship cannot wait. It must be rushed down the ways to the open sea. For the world is at war.

They are so small against the enormous hull, these men who have shaped it. Drained of vitality by the creature of their own making, they seem of little value. This mountain of steel that blocks the sky is mightier than themselves. They have formed something beyond their comprehension. A welder, pausing to wipe the sweat from his eyes, looks up at the great ship with vague, inarticulate pride. He follows it in his mind over the sea, watching it battle against enemy and storm, listening to the boom of the guns. But looking at it thus, in awe, he fails to realize that this ship they are building is controlled by man.

The same sun that blazes upon the shipyard, heating the steel until it blisters the men's hands, falls upon corn fields and wheatlands across the surface of the earth; and as it touches the corn and the wheat, it draws up into the plant from out of the earth power for growth. But the battleship has no power for growth. It is built of cell-less steel, a creature alien to sun and earth, light and rain. No sap rises into this body at the touch of the sun, and the great bulk holds less magic than the smallest field mouse who is delivered of her young in a nest among the stubble.

Within the main gates of the shipyard stands a pile of rubbish. It is part of the foundations that were dug when the shipyard had been enlarged. Rusted wire is coiled around broken bottles and old tin cans. Scraps of torn paper have been lodged here by the wind. But there are one or two little pockets of earth, between the rubble, and in one of these a dandelion plant has found roothold. Its leaves are greyed with dust, its flowers battered by trampling feet. But it has lived its appointed cycle, and the seed-clock stands erect, awaiting the breeze. And then, as the ant-like humans weld and rivet the great ship, a little breeze rises, unnoticed by the sweating men. It crosses the shipyard and reaches the pile of rubble, where it lifts the seeds of the waiting dandelion plant and carries them through the air in its passage. Over the heads of the workers floats the little parachute, a tiny brown seed at the base of the dandelion down. It is so small that it is not observed. It is so silent that it could not be heard, even were this pandemonium of the shipyard suddenly stilled. Over the great steel ship floats the dandelion seed, somewhere to fall into the earth unnoticed. But as it sinks to earth it will have carried within its brown shell the possibility of immortality. For it holds the power for birth and growth, and descendants of this one little parachute, that floated across a shipyard on a summer day in wartime America, may root and bud and blossom and seed when the great steel ship has fallen asunder and the great-grandchildren of the shipbuilders are waiting to be born.

Man, who can build ships and span the rivers with bridges, does not rule the world. The seed that is dropped into the earth is master of millionaire and banker, riveter and politician. It controls worker and king. For the seed has the power to grow, and without it we must starve and die.

Out beyond the city, a farmer's wife rests on her porch. Bewildered by a world at war, she lifts her eyes to the fields around her.

"And yet it's all so simple," she murmurs. "If only people lived here in the country they'd know it was simple, for they'd see that the one thing that matters most in the whole world is the earth. Everything we eat comes from the earth. And you've got to eat to be able to work, or drive a tank, or make an airplane. You don't care if you live or die, when you are starving. And if you don't care what happens to you, you don't even want to build ships, and tanks, and guns, much less make a new world when this war is over. Somehow, it seems as if life had gone all wrong, for there it is, so simple and clear, if you'd only use your eyes."

She pauses, looking towards the farmer who is coming in from the barn. Now she goes on talking.

"There's some that think only of the money they're making out of this war. But money isn't any good to you if there's nothing for you to buy with it. You can't eat dimes, and you'd only choke on the paper of dollar bills. . . ."

She shakes her head, and gazes out over the fields. Before us lies sleek farmland, yielding corn and oats and hay. But even while we look upon this peaceful scene, the calm is broken by loud explosions from the army proving ground, over the hills. The structure of the porch shakes beneath us, and as these thunderings break into her talk, the farmer's wife turns her head from me, towards the hills; for her sons are fighting, and her daughters have left the farm to work in those plants that thunder across the land.

Sitting here upon her porch, we see, as in a vision, men and women working at lathes. We see them making shells and airplanes. We see them standing at production lines, dazed with fatigue as they turn out the weapons of annihilation. And sitting here together, we two women, we know that we live in an age that runs counter to the ways of woman. For though man is the fighter, woman should be the creator and preserver of life. Now, at this moment, we are bound to each other in our distress; for we face a world that has been forced into the path of destruction. And with this moment of insight, we know that we must continue to weave the central pattern of life, that the thread of civilization may not snap. We know that we hold in trust the things that must endure. Over in the war plants, the workers stand before their machines. But behind these machines lies the earth. Unspectacular and quietly eternal, the earth is the fount of all that muscular power. It is behind the force of the Bessemer converters at the great steel mills, and it drives the propellers of the bombing plane. It welds the plates of the battleship, and supplies energy for the man who fires the gun. Undisputed in time of peace, still it is supreme in time of war. For behind each army that fights, and behind each brain that plans the design for a lasting peace, lies a field of corn. Somewhere a farmer planted gold kernels of corn into the body of his earth, that hogs and cattle might be fed; and somewhere far removed from this field, human minds, nourished upon the harvest of this corn, decide the destiny of our world.

It is little wonder that the farmer's wife, with the wisdom of all simple people, knows that she and her fellow-workers upon the earth shape humanity.

## 1: THE ARTERIES OF SUPPLY

The sun beats down upon another scene. It strikes the metal of derricks and cranes, till they burn hot like molten steel. It is reflected in the buckets of travelling belts, and mirrored in steam shovels. It shines upon the sweating torsos of work gangs, and lies in broken bands of gold upon the water.

For a ship has docked, and her cargo of coffee must be unloaded.

Insignificant against the great bulk of the cargo vessel as the welders were against the battleship, these dockers weave their human thread into this involved pattern of gravity chutes and cars, travelling belts and cranes. It is a scene of bustle and noise, and the mewing of the seagulls in the sky above the ship's masts is drowned by the rattle of derricks and chains.

This scene we watch will endure long after the return of peace. For its substance is food for the cities, so that men may work and think and live. Here to the docks come fruit ships and grain ships, and here by the water's edge stand cold storage warehouses and the box cars of freight trains. Auctions and terminal markets are grouped around depots and docks, for behind the life of a city pulsate the arteries of its food supply. Into the city come cargo ships, freight trains and trucks, that the people may be fed. For the drama of food is staged not only on corn field and farm. Out from the earth, which is the heart of the world's body, comes all that we eat; but as the heart pumps the blood through arteries and veins, that it may reach fingertip and foot, lung and brain, so must the arteries of our food carry this lifeblood of milk and meat and fruit to the far corners of our civilization. The lungs of our country, the brain-centers and finger-tips of our country—city and port, university and mining town—into them must be pumped, from cattle ranch and farmland, the blood stream of food. Across the water, along the railroad tracks, up and down the highways that blood stream flows, its course complicated by commission men and brokers. But this life-blood of food forces its way through freight rates and bills of lading, markets and chain stores, to the table of the people.

States far-distant over America meet upon the breakfast table of an insurance clerk in the Bronx. As they eat, this clerk and his family, they are fed from earth they have never seen, and nourished by trees and rivers that are no more than names to them. Out of the electric toaster springs bread that was bought at a chain store around the corner; but, beneath the cellophane covering, this bread that has been baked in New York is made from flour that was ground in Buffalo. This flour came eastwards in a ship along the Great Lakes; it came eastwards in a freight train, as hard wheat, from the Dakotas, where it was "combined" upon a field that met the sky. The bread they eat, this family in the Bronx, has been mixed with lard that may have been refined in Omaha, from hogs that were shipped from a farm in Nebraska. Upon this bread the family spreads butter that has come from Wisconsin creameries, churned from the milk of cows that grazed near Madison. Oranges rumbled across North America in freight cars, to bring fruit juice from California, and the tropical heat of Brazil ripened the little magenta berries from which the coffee is made.

The wide-flung countryside showers its food upon this breakfast table in the Bronx. The insurance clerk lifts his hand to the sugar bowl, and trucks are loaded with sugar cane upon a plantation in Louisiana. Chick hatcheries were at work months before this, that he might eat his egg; and the corn that fattened the hog from which his bacon was cured had been tossed into a pen in Iowa.

Beyond the railroad tracks and the highways of America, out past the coastline of this country, people labor for our food. Colored gangs load the fruit ships of the tropics, singing strange chants as they work. Across the ocean comes tea, its leaves gathered by alien hands.

All these harvests of wheatland and corn belt, orchard, dairy farm and plantation, come together to supply the stuff of vitality to one small family in New York.

A sick woman lies sleepless in a bedroom in Illinois. The wind skims the corn fields, and its murmur is like the swish of harsh silk. The woman wonders what time it is, for she seems to have listened to the wind in the corn through many hours. And then, as she turns restlessly in her bed, she hears a comforting sound. It is the dull rumble of the nightly freight train on the Sante Fé. With the relief of the sleepless who had dreaded the long hours until dawn, she hears the clang of the bell, as the train nears the grade crossing; for she knows now that the dawn is near. Twenty minutes to four o'clock in the morning, and the freight cars emerge from the blanket of night, closer and louder until they roll past the far edge of the corn field, car after car after car, as though their procession would never end. If it were daylight, the sleepless woman could have counted ninety-seven cars, but now, in the blackness, she listens for the moment when the rumble shall lessen until the wind carries it away and folds it into the night. Soon the red glow from the engine will fade, for the ninety-seven cars will have passed across her world, and all they will have meant to her is the comforting assurance that it is nearly four o'clock.

In the caboose of the freight train a bunch of cattle hands plays poker. Their baby beeves in the cars of this train are far from their minds. Above the rattle of the train the men shout, and the drowsy brakeman is jerked awake by their noise. The stars above Illinois look down upon the freight train, playing a game of hide-and-go-seek with the smoke from the stack, and the tail lights are like red and green stars that have fallen from the sky, and been hitched to the caboose.

Over the countryside rattles the train, a long chain of dull windowless red box cars filled with grain, double-deckers filled with hogs, yellow refrigerator cars filled with fruit and vegetables. Inside three of these yellow cars are oranges from California. They have crossed the fierce desert behind the coast, but never has sun touched them, or heat shrivelled them, for they are secure within the iced cars. Over the Rockies they have travelled, across a Continental Divide, with locomotives added, and the freight train cut down in size, for the upgrade of the mountains. Past the Rockies to the plains of Kansas, and the cars of oranges from California, the cars of apricots, peaches, cantaloupes and lettuce from the Imperial Valley, have been joined by more cars, now that the load is pulled with greater ease across flat country; cattle from Texas, and corn-fed hogs for the meat market of Chicago; grain for the Great Lakes ports; eggs, cheese and butter for the cold-storage warehouses of Chicago.

The oranges from California are bound for New York. Across the states of America they move eastward, and automobiles and trucks wait with impatience at grade crossings, to let the never-ending train pass by. This freight train holds within its tedious length the drama of transportation. It stands for the great impersonal plantings, harvests, and markets of today. A farmer's wife in New York State pauses to look towards the railroad track in the heat of the summer afternoon, as the freight train rattles by. She leans upon her hoe, for she is weeding lettuce in her yard. But within those dull yellow cars is more lettuce than she has ever dreamed of; crated lettuce from California for the salads of city dwellers, lettuce for hurried salesmen who eat their lunches at the counters of corner drug stores, salad for the people who have never seen lettuce grow, or smeared their fingers with the milky juice from its snapped root.

For the cities must be fed.

But the pattern of the feeding of cities is not simple. Behind each sandwich in the corner drug store is a chain of freight cars and trucks, docks and men, of grain elevators and cold storage, of packing houses and display pens. Freight train moving across America; stevedore unloading cargo at the docks of the Great Lakes grain ports; truck driver hauling his load of canned tomatoes, drunk with need of sleep at the day's end: all these have served the sandwich in the city drug store.

The sick woman lying sleepless in Illinois is not alone as she listens at night to the food of America flowing across the country. In an old stone inn on U.S. 40, the spirits of pioneer women turn in their sleep, as strange noises pass upon the National Pike. Their phantom husbands, resting beside the covered wagons in the meadows opposite the inn, wonder at these horseless creatures tearing over the Old Trail. Along that National Old Trail, reborn as U.S. 40, trucks rush west and east. They overtake the slow spectres of creaking prairie schooners, colliding with them upon the empty air. They are filled with foods unimagined by these pioneers, for cities that were unbuilt in the pioneers' day. They rush west, from Baltimore and Philadelphia, with canned goods from the Coastal Plain. They rush east with dairy produce from Ohio and apples from the Shenandoah Valley, for the cities of the Atlantic Seaboard. They bring sugar and coffee from the ships of the Gulf Ports. Over the heat of the prairie they go, and the long straight road that meets the sky is a mirage of water. Up the high mountains they crawl, switchbacking across the valleys to further mountains, through smoke-dimmed mining towns and cities of steel mills, past fields of tasseled corn; and upon their heated bodies lie pollen from the deserts of the west, and bird droppings from the Rockies.

The sun sinks behind the Appalachians, and darkness falls. By the side of the road, along U.S. 40, some trucks are parked for the night; within the cabs, companionless drivers sleep till dawn. But into their heavy, dreamless sleep breaks the familiar roar of other traffic that passes even in the night. Ten hours at the wheel of one of those ever-moving carriers, and another driver has wakened his mate, asleep upon a bunk in the cab of the truck. For these there is no pause; and the truck drives east, through the hours of night. The stars that look down upon the freight train in Illinois look down, too, upon yellow headlights devouring the highways of America.

Through the cold of winter and the heat of summer, battling against snow-drift and blazing sun, ice and storm, truck drivers cross the American land, bringing food to the cities.

The drama of our food supply is being played while most of the people in the cities sleep. But the late-shift war workers can watch it. As they stumble home to bed, returning from shipbuilding yards and airplane factories, they see trucks converging toward the produce market in the downtown district. It is one o'clock at night, and the auction begins. In from the darkness come truck-loads of vegetables, and it seems strange that these green things of daylight and sun should be out in the night, like children who have passed their bedtime. But the sellers are not worried by such thoughts, and with the speed of all auctions the vegetables change hands. When morning comes they will lie displayed in the food stores of the city, to be bought by housewives who are now asleep.

The nocturnal commotion shifts. It is the fruit exchange now that is filled with buyers and sellers. They crowd the railroad cars in the terminal sidings, comparing the crates of oranges and grapefruit. Perfect in color and shape, the fruit of this market is checked and sold with hurried deliberation. Here is no place to pause and dream of the beauty of orange groves. Men exchange this fruit for dollar bills, and any magic that is here is far from the earth. It is the magic of the gamble of the market, and the counters in the game have the names of the types of sales: Consignment, F.O.B. Sale by Contract. The men who play this game are the "fobbers," shippers, brokers, commission men. All these concern themselves with an orange, all these and the men who load and pack, and the men who sit at desks to make out the lists of freight rates and demurrage charges. The one thing that is far from all the clutter of the market is the earth; and it is the one element without which none of this activity could exist.

Man is proud of his efficiency today. He can look with pleasure upon these arteries of supply, knowing that he has vanquished distance and subdued time. A civilization that can convey milk across a continent, and send it to a breakfast table fresh as when it came from the cow, is justified in this pride. But mechanization is charged with danger. Man is still young with his machine, and he is unaware of its subtle power. For he is not a robot. Starvation of the human spirit is as terrible as the more visible starvation of the human body. Only to a certain point can man dare to be impersonal. Unconsciously he needs human warmth, and closer touch with his fellow men. For the human drama is played invisibly, within the hearts of men; and competence and efficiency are not enough. We dare not lose delight in life, and the sense of wonder. So, even while we watch with awe the co-ordination of our arteries of supply, we see workers inarticulately hungering for comradeship. We see men caught into the treadmill of mechanization, who are servants of the machine rather than masters, and who lack, in their exhaustion, those moments of magic which give meaning to life.

But a benignant ferment still works. All intense emotion reaches beyond the radius of its actual existence, and one small market place of unashamed human warmth, where the people sit in beauty among the plants they have grown and the chickens they have tended, holds the power to send this warmth into the world, to those who have forgotten that man is greater than the civilization he has created.

In a city in Pennsylvania stands such a market. Within an ordinary red brick building is housed a serenity that is strangely sane against the madness of our world. Into this building are gathered the fruits of the rich farmlands of Lancaster County; the fruits of a creed of ceaseless work upon the earth and the harvest of tender care. But there is something to be felt here beyond this. It is the awareness of beauty. Enter these doors and you come into a world that dares to be eternal, and is beyond fashion. It is a biblical world, with market people who might belong to the Old Testament. These

Amish and Mennonite farmers keep hold on serenity in the midst of the confusion of the market, for they have their peace within them.

It is strange to see these people against the background of a city, for they seem to belong so completely to their fields and farms. But they have the power to overcome the fret of sidewalks and crowds, and, as in the little markets of Europe the peasants brought into the turmoil of the town the spirit of their mountains and valleys, so these bearded Amish farmers and bonneted Amish women bring with them the color of ripening barley and the scent of clover fields in flower. And this is what the city needs, as much as it needs sustenance for the body. The bustling crowds heap into their market baskets something a little more than butter and eggs: it is the tenderness these farmers felt for their cows and their hens that passes from seller to buyer.

The people mill around in a confusing pattern of movement. But though the Amish farmers are always busy, and never still, they carry the quietness of creatures who know how to rest. That tall man at the stall towards the door has a biblical peace within him more fundamental than the Old Testament beard he wears. As he rearranges the bunches of young carrots and the spring onions, his enormous hands touch them tenderly, as though he were settling them in their soil. And then it is, as we watch this tall farmer with the gentle eyes, that we see what has happened. These Amish and Mennonites have brought their earth in here to the market building with them. It is so much a part of their lives that they are unable to separate themselves from it, and their eyes still see the stretching fields, disced and harrowed for the planting of corn, and watch the cows that graze in the meadows near the buttercups. Elias Redgay, the tall farmer with the beard, turns to the bunch of buttercups in a jam jar at the far end of the stall; the flowers are fresh, still, in the heat of the market, and shiny and polished as his wife's face, or as the scrubbed vegetables he has arranged in decorative clumps. The philosophy of these people can be discerned in the presence of this bunch of buttercups. For they need beauty, and no one considers that buttercups are merely weeds that grow untended in the meadow near the stream. This golden bunch is a token of the honesty of their hearts.

Flowers are scattered everywhere in the market building. The people who sell eggs and chickens have turned their stalls into bowers of bloom, until there is the feeling of harvest festival in a village church. And perhaps the Amish farmers have the rare wisdom that makes little difference between their God and their earth. As they have been known to hold religious worship within the shelter of the great barns, so here, in this market, a sense of worship surrounds the fruits of their earth. Austere in their own lives, in their love for the land joy and color are allowed to break through.

The beauty of this Amish market, seen against the background of war, is such that it touches the heart. And the beauty that we feel does not lie merely in this strange sense of abundance in a time of scarcity. It goes deeper. It is a beauty that is entirely right; for it satisfies a need. This need arises out of the loneliness of human beings. Man lacks something when he is forced to buy his food with the impersonal attitude of a motorist at a casual filling station. He needs to belong with his fellows, and is impersonal at his own cost. Something is lacking to the shipbuilder who does not even know the name that will be given to the ship he rivets. Something is lacking to the salesman in the chain store who does not know the name of the housewife to whom he sells butter. We carry with us a loneliness that we fail to understand.

In the Amish market, the people buy food that has been raised by neighbors, upon native earth. The crowds who shop in the impersonal markets of the world may not be conscious of their own need; but let them look at the contentment in the eyes of the Mennonite farmers, and they would be convinced of what they lack. Rachel Harnish, the young girl in the pink flowered dress, with fair hair drawn smoothly back from her face, and cheeks that are almost the color of the dress she wears, has helped her mother bake those pies she sells, and has grown and gathered the flowers that adorn her stall. Something of this has entered into the tranquillity and beauty of her face.

Outside the building is a confused world; a world that seems sometimes, despite its efficiency, as though it had lost its way. Within the building we have seen beauty and sanity. For we have seen people who revere the foods we eat, until for them and for us corn and wheat, meat, vegetables and fruit, milk and eggs, hold an aura of magic, and the land upon which these grow, and the farmers who raise them, are deemed worthy of praise.

## 2: THE FARMER

The freight train crosses Illinois. It comes out of the West, against an orange sky, and as the sun sinks below the horizon the land darkens and the train passes great stretches of blackness; for it crosses the vast fields of the Corn Belt. But from time to time it comes to a cluster of lights. It slows down at a grade crossing, and the shriek of its whistle is tossed back from a grain elevator that dominates a little town. As the train slows, the engineer leans from the locomotive and sees crowds moving upon the sidewalks. He sees farmers and their families; for it is Saturday night.

It is Saturday night in a little grain town in Illinois. The streets are filled with cars, and the shops are thronged with farmers' wives, buying their weekly provisions. Children crowd the corner drug store for ice cream cones, and old women sit in the parked cars, rocking their daughters' babies to sleep. And while the wives market, the farmers tread upon the sidewalks with the lumbering steps of men who are accustomed to the rough earth of corn fields. Their faces and arms are dark from the sun, and when they remove their hats, to wipe the sweat from their heads, the skin of their foreheads is as though whitened with chalk. They stroll from group to group, and the words we overhear are the words of their trade.

"How are you making?" asks a gentle-faced man in clean blue overalls. And the answers that reach us are: "I finished my beans today," or "He's got too much land to work properly," or a low mumble that ends with "a row as crooked as hell." From nearby groups come words like "team"; "acres"; "corn." For the farmer lives with his earth.

As the womenfolk with their children wander along the sidewalks in their summer dresses, this quiet little town seems changed suddenly into a garden of flowers, with blossoms swaying in a breeze. There is beauty here, for these people have brought into the town the spirit of their fields. But the gods have been kind to them, too, for they are rugged and fair. They stand around in gaily colored clumps, illumined by the lights from the shops. One thought is in their minds, and if their eyes search the skies, it is not to notice Venus, high above the grain elevators of the town: it is to wonder how soon the screen will be hoisted against the blank side of a store, across the railroad track, for the weekly movie show.

Night deepens on a corn town in Illinois. June-bugs bump into the faces of the farmers and their families who crowd the tracks and sit on the cross-ties of the railroad. They sit here to watch the weekly show of life upon a screen. Ears that have heard the plaintive cry of the kildeer, and eyes that have seen the miracle of birth and growth, listen to the music of Hollywood, and gaze upon the synthetic intrigue of the films.

But reality sends the freight train along the Illinois Central. Into the noise of the movie is woven the distant rumble of the train. Above the noise sounds the piercing whistle of the locomotive, and the people jump from their seats upon the cross-ties of the railroad, as the freight train approaches. Through this little town that was so recently unbroken prairie, the train rattles. It carries grain from the wheatlands, corn from the great fields of Illinois, cattle and hogs for the packers of Chicago. The train that interrupts the drama upon the screen bears across the land the essence of these farmers' lives.

But who are these farmers who produce the grain that fills the cars of this freight train? What are they like, the men who raise these cattle and hogs, and harvest this corn? And what is the rhythm of their days? For their lives are spent in working, and it is not always Saturday night.

The black loam is still water-soaked. You can smell it as the tractors plow the furrows. For the farm lies low, and the floods this year had turned the fields into lakes. It is the middle week of June, and the land is unbroken still, and tangled with weeds. Three hundred and twenty acres of good Illinois earth stand unplanted and unplowed; and the world at war needs food.

But something stronger than flood has broken this farmer. Something of more force than the ceaseless rains has checked his wife in the planting of her garden. Phantoms stalk their fields, standing between them and their sleep at night, and between them and their land by day. This farm is dimmed by the spectres of war, and the spirit of its owners has been defeated. For the three tall sons who were born upon this land, who plowed this black loam and planted and harvested the corn, have gone to war.

The land has conquered this farmer and his wife as they work alone. Happiness has left them, and with tired bodies they try still to battle against the earth.

But something is happening this morning. Along the straight roads of Illinois, and around the curves of the township jogs, move seven tractors. They have plows and harrows attached to them. They are followed by a four-row corn-planter, and a truck loaded with sacks of seed corn. This mechanized farm army converges upon the straight roads, to the low-lying farm that battles with the solitary farmer.

The tractors are driven by neighboring farmers. They have come to help a fellow in distress. Shyly these men explain away their generosity, as though they were ashamed. "As like as not it might have been that the floods had been sent to one of us," mutters one man. "It'd be a waste to let all this land lie idle when the world is wanting food so badly," says another. "And it isn't as though his three sons weren't fighting this war for all of us," says a third. The farmer who has brought sacks of seed corn murmurs that he had miscalculated in buying his hybrid corn this season, and is glad to find a use for it.

They have opened up the first furrow, and the smell of oil and gasoline from the tractors blends with the smell of wet earth and crushed green weeds. The turned earth is black and shiny, as the green of grass and weeds is laid low. "This farm is sure not one man's job," think the men, as seven tractor-drawn plows break these fields.

On the wooden steps of the porch the farmer's wife stands dazed. She can scarcely believe what she sees here before her. She watches her man at the far end of the field, driving his own shabby tractor with the two-share plow, and it is as though vigor had returned to him. Seven neighbors have given him something more than the planting of his fields; they have restored his waning courage.

"Simon and me, we'll sleep mighty tonight," thinks the farmer's wife. And she turns to the house to search in a drawer for packets of vegetable seeds. The seven bright tractors have planted the spirit of courage in her own earth.

Throughout the long June day men plow and disc and harrow and plant a neighbor's fields, and the sun sets upon three hundred and twenty acres of good Illinois earth, within whose black depths lie rows of seed corn.

The true farmer has a feeling of dedication to the earth, which transcends ownership; mankind is one large family, and fields belong to all men. Let a farmer be the victim of the elements, his neighbors will be there to help him. For the farmer knows there is a force beyond his control that can destroy him, and this binds him to all who work the land. "It's a powerful good thing we can't have a say in what the weather does," says a German Lutheran farmer in Maryland, "for we'd never be able to please everyone of us. I went and cut my hay yesterday, and today comes a heavy rain, and as likely as not it will be ruined. But if I'd been able to stop the rain, perhaps some other farmer would have been needing it for his hay to grow. And when the lightning struck my barn and burnt it to the ground, with my year's grain in it, and the fodder for the cattle, and I hardly had time to get the animals out to safety—why, I knew for sure it wasn't the work of someone who wanted to do the dirty by me. . . . It gives you a feeling we are all being treated alike by some outside power, and we'd just better help each other when it strikes."

This feeling inspired seven men to plow a neighbor's fields in Illinois.

But though the true farmer has this communal sense about the earth, it does not mean that he lacks personal love for his own land. And this love is needed. It is something of value that we have brought with us from the Old World, for it comes from a civilization where land was less abundant, and consequently more precious. The French peasant with his little parcel of earth loves this earth with a passion that cannot belong by nature to the farmer of the Middle West, whose acres run into the thousands. A new country like America, where man might plunder the soil without scruple, because always there would be untilled land out West, could not develop within the spirit of the pioneer a sense of identification with a particular piece of earth. Were his land to become depleted, carelessly he could move to richer land. But America today is frontierless. There is no new land to plunder, and something of true care for the earth inevitably will grow in the worker upon the soil. Re-pioneering lies before us, and the reconstruction of raped earth will force man to cherish this land. Man loves what he serves, rather than what he ravages. Remembering this, we understand what makes the Mennonite farmer a magician upon his land. He has a sense of service to the earth, knowing he holds it in trust for his sons, even as he himself has received it from his ancestors. And so as we look upon the rich, cared-for farmlands of Lancaster County, Pennsylvania, or the wide lands of the Mennonite communities in Central Illinois, we see evidence of their faith and know that they hold this land most dear. For more is needed than good plowing and fertilizing. Something other than tractors and combines must enter into the tending of the land. It is not enough that we should boast of the fertility of our soil and of the volume of our crops. We need to love the soil, and to love it with a personal devotion. Insensitive ownership robs the land of that tender care which alone can make it flourish. We could learn from the peasant of the Mediterranean whose terraced fertility has been wrested from the mountain sides, from the Italian immigrant who cherishes his grape vines in the Hudson Valley, or from the fat little Czech farmer's wife who works her piece of land in Western Pennsylvania with the same affection her ancestors had shown for their farm in Europe. Above all, we should watch the farmers over here with the German names: Amish, Lutherans, Middle West farmers whose ancestors left their fatherland nearly a century ago, seeking freedom and cheap land. It is here we can learn our lesson. These are the men who belong with the earth they till. These are the people who feel unashamed sentiment towards their earth. For we must mate the enlightenment of science with the emotion of the peasant, and harvest from the land, as well as soybeans and corn, mythology and song.

Man who tills the earth needs the companionship of the gods. But they may be new gods. He cannot adopt the mythology of the Majorcan farmer who winnows his beans upon the circular floor of the threshing ground to the songs of his forebears. It is not enough to praise Bacchus with the Provençal peasant in the vineyards of the Mediterranean hillsides. Man in this New World must have his own dreams and his own new visions. Upon the wheatlands of Kansas, on the cornfields of Iowa and Illinois, in the dairy farms of Wisconsin, there are new gods awaiting discovery. They are the gods that haunt the combines and the scarlet cornpickers. They live in the solemn shapes of silos, and sing their songs in the windmills of artesian wells upon Texan prairies. They come in the guise of dust storm and chinook, as surely as they visited the earth of the Old World as lightning and thunder. There are patron saints waiting to protect cattle roaming the open range, and spirits in the orange trees of Florida as full of power as sister spirits in a thousand-year-old olive tree on an island off Spain.

But, also, there are farmers who have an awareness of the divinity within themselves. William Schuler works his four-section farm in Illinois with the devotion of one who feels that he labors with God. Driving the tractor across his four-hundred-and-forty-acre field, for the planting of his corn, he knows he is a creator. Knowing this, he shares the humility of all truly creative persons, whether it be scientist, composer, or poet. And how could he but be humble, he who watches the miracle of the sprouting of his corn or the farrowing of his sows?

William Schuler sits at night in his parlor resting. This tall Illinois farmer, with the gangling figure of Abraham Lincoln, sits talking with his hands—the huge hands of a farmer that seem to be feeling the moisture in the earth or the milk within the ear of corn even while they are empty of earth or corn. As he talks, the light of the visionary fires his eyes. "All men are created equal and free," he says, opening the palms of those great hands. "My six hired men are free, and they know it. Their only allegiance is to our earth, and it is their earth they work, as well as it is mine." He turns then, to show us the picture of a favourite work horse. "She's twenty-two years old, she is. And she's just like my hired men, who've been with me about as long as she has. She's just like them because she's well kept and because she has confidence and good feed. That's what the whole world needs, when you come to think of it: confidence and good feed."

Looking at this Middle West farmer as he talks, we know that beyond his store of scientific knowledge, he carries within him instinctive feelings that are deeper than knowledge. For he carries within him the sense that respect is due to all men, and to the whims of all men. It is this respect that makes him raise a field of sweet corn, so that his workers, who came mainly from Kentucky, may not lack the accustomed cornbread of their youth. And this same sense makes him delight in showing us a photograph of the laughing children of these hired men. Here, we feel, as we look upon this farmer, is what America stands for. And it is as though he read the thought within our minds, for he goes on talking to us:

"We need to be allowed to be free. That's what we all came over here for. And freedom doesn't need to be destructive. It can't be, if you work in the earth, for you are in bondage to that earth always and everywhere. But if a man feels the need to serve—which he does—what better master could he find than the earth? . . . Farming should be something you love. But take away the farmer's initiative, and you make farming into the worst drudgery a man could carry. That would take the whole of our happiness from us."

Listening to this man so, we see how completely the farmer is bound to his earth. And suddenly the great form of William Schuler fades, and before our eyes sits a worn little man nearing eighty. He has sat in this chair for more than fifteen years, crippled with arthritis. But in his mind he walks his fields and plants his corn, as he had done for half a century. His distorted hands hold a black notebook. Within this notebook stands the record of each day since the year 1900: March, with "We have nigh Spring, soon will plow"; April with "Butiful Warm and Growing weather, Lookes for Rain"; December with "Fogg, lowery day." Here with red lines, upon a map of his farmlands, he has traced the course of the new eight-inch tiling of the East Forty, and now, from his chair, he plants willows in a ditch at the start of the century, that his soil may not wash away; and his knotty hands grow tired with the labor of making a pond where the ditch can drain, to be used as a hog wallow. Looking at the rain beating against his window, he shakes his head as he says it is the wettest spring since 1908. For his life has been controlled by the elements, and his father's life before him. Samuel Keller stretches out an unsteady hand for another notebook, shabbier than the first. It is dated 1860, and within the faded brown ink lies the epic of a pioneer farmer. For it tells the story of hard living: "Mother and I have bin verry saving we aught to have something to show I feel the pressure often but wont give upp." Behind these lines is hidden the greatness of America. Old Samuel Keller knows his own worth, for he knows he is the son of a pioneer. And when his pain seems more than he can bear, he turns for sustenance to this devotional book, the diary of his father. Here, on these dimmed pages, his father talks to him. "February 12, 1894. I walked out to the Barn the first since early January about the longest time for me to be Housed. I feel I have found a friend agin. . . . June 19, We at hay. Got a grait bigg stack. We are verry tired. I could hardly stand the heat." And should Samuel Keller feel he has failed this pioneer father of his, he turns to a much worn page in the beginning of the notebook, on which is written:

"This is grait weather to work in the Corn. Fine son was born, weighing 8 pounds. We hope he will grow to have the moral Worth of his brother Asa."

Samuel Keller sits in his invalid chair, but he is of the black loam of Illinois, as was his father before him. So completely is he one with the earth of his farm that it is almost as though the milky corn grew out of the man himself, here in the back room of a house in town.

Out upon Samuel Keller's farm today a seven-year-old child rides in glory; for it is hay harvest, and as the hay fork in the barn is pulled by an old grey dobbin, little Sammie, the great-grandson of the man in the invalid chair, has come into his own: he is old enough this year to ride the horse. His tiny limbs can scarcely stretch across the creature's broad girth, and the child feels himself slipping upon the sweat of the animal's back. But fear adds to excitement, and this, today, is the initiation of the youngest Samuel Keller, as terrifying in its own way as the initiation rites of Australasian aborigines. Soon he will be allowed to take water to the thirsty harvesters, in great earthenware jugs with corn-cob stoppers. He will be sent by his mother to call his father to dinner. If he had happened to be born in the South, he would have started his novitiate with the chopping of cotton; but here in the Middle West he will help with haymaking and harvest. By the time he is twelve years old he will ride the harrow, and, he tells himself, they do say a boy of twelve could drive a tractor. . . . The old grey dobbin carries a burden worthy of the name he bears. For this small creature is already part of the black earth of Illinois. The great-great-grandfather who first broke this prairie land brought west with him, in the covered wagon, plow-share and hoe. If his spirit still haunts these fields, bewildered by the strange forms of tractor and corn-picker, he will surely look with pride upon this youngest member of his blood, knowing he did not break the prairie in vain. History is still in the making upon this farm in Illinois.

History is being made, too, by the woman on a farm. Behind the food stands the wife. Behind warmth in winter, shade in summer, comfort in time of distress, stands the woman. The pattern of her days is filled with ceaseless working and planning, and when her menfolk return at sundown, their limbs shaken from a day with the tractor, she mends the overalls that got torn upon a nail and darns the socks that are rubbed into holes from trudging the fields. The farmer rests in the evening, smoking his pipe, but the woman must mend and bake for tomorrow.

Her day never ends. There is always a baby to tend, or hot mash to prepare for a sick animal. Washday means scrubbing the men's overalls that are caked with dried earth, greasy with oil from the machinery, bloodstained from the butchering; the yard is festooned with lines of overalls and shirts. And in the scandalous heat of a summer afternoon the smell of the hot iron catches the throat. . . . The harvest hands will be here in a few days; she will have to rush home from riding the binder, to cook for them—hot breads and meats, and vegetables from the garden. How ravenous they are, when they come to the house at noon! They have cleared the dishes of chicken and pork and lamb before she has time herself to sit down and eat. And harvesting always comes when she is canning her beans. . . . There was that summer of the drought, too, when the heat struck through you like a knife, and little Rufus was born; try as she might, the babies would be born at harvest time, when she was needed in the fields. Most of them had come just then—Rufus and Dot and Connie and Bill and Tom—and when it wasn't then, it would be at the butchering. Emmie was born on the actual day, and if the minister hadn't helped with the butchering, what would have happened, you just can't imagine.

The hired girl comes in from the milking. There is supper to set. . . . The sun is low in the sky and the farmer's wife must water her plants. She goes to the front porch, to tend the dozens of potted plants there: begonias, geraniums, ferns, cactus; she knows the exact shape of each plant. Tenderness comes into her face as she stoops to water a gloxinia: twenty-two flowers it had last year, the color of a ripe strawberry when you slice it in half. She works among her plants, and there is a softness in her movements, as though she would stroke each leaf and flower. The mother in her that has raised eight children, the warmth in her that tends the menfolk and the garden and the chickens, finds in these plants something more to love, as though there were no limit to her need of life to cherish. She moves to the back porch, where she has as many more plants to tend. She walks through the house, and in every window grow creeping plants, hanging plants, budding plants that must be cared for.

As she stands here in her kitchen, stirring a pan upon the stove, she seems a worthy figure of womanhood. Against a world given over to destruction, she symbolizes the preserver of life. There is in her something of the kindly abundance of all earth growth and earth power. She is beyond change, and will endure as the land endures, and daytime and night. Hers is a life without hates, that is big enough to care for the fields of other farmers and the children of other mothers. Man in his innermost lack of security may still need to talk of "my field" and "my harvest," but deep within him the good farmer can think in terms of "our fields" and "our harvests," and his woman can fill her arms with the care of all men and all earth. It is this that glows in the face of the farmer's wife and gives her the sense of quiet, inarticulate power.

## 3: CORN AND WHEAT

AMERICA IS CORN. IT WAS BUILT ON corn, and exists still upon corn. The freight train that comes out of the Middle West is the product of corn, for it carries within its cars cattle and hogs, fed and fattened on corn. The stockyards of Chicago, the hog farms of Illinois, the dairies of Wisconsin, exist through corn. White-faced yearlings in feed lots over the country are fattened on corn. Belted Hampshires gain weight on corn. Rows of Guernseys and Holsteins in stanchions across America are milked to the soft sound of the munching of corn: for the feed troughs are filled from the silos.

Corn is pork and bacon, spare-rib and sausage. It is steak and hamburger, milk, butter, and cheese. It is eggs for the people of America. It is cornbread for the man in the South. It is man himself, and the spirit of the New World; for it stands for the growth of a civilization, with railroads and freight trains, stockyards, markets, and cities. The men who sit behind desks and work at the tills of banks are the outcome of a corn culture, and the cities with their towering skyscrapers should pay homage to the corn.

Back over the ages, the forerunners of these city men lost their liberty to a kernel of corn. With the planting of this kernel they offered hostage to the earth, and surrendered their freedom to roam. Tamed to the sprouting grain, their lives were bounded by the cycle of the earth's year, and the nomad spirit of the hunter died.

As the forebears of these city dwellers were tamed to their growing corn, new fears came upon them. Skies that had been of little significance while men lived still by the hunt became the focus of their eyes. For upon the skies depended the food they ate. And with these new fears came new beliefs and new gods. Maize magic demanded that tribute be paid, and sacrifice offered, to the gods of fertility; for without their goodwill mankind would starve. Men found ways to please the Earth Mother, that the seed might sprout, and danced their dances to the rain gods, that the maize might not wither upon the stalk. And the people who had lived recklessly by the hunt became tamed to a new pattern; secured to the tending of their fields, they gathered around them a family, and civilization developed, dominated by a seed of corn.

The sacks of seed corn stand today at the end of the barn. These sacks hold the power to change the face of the countryside outside this barn. The alchemy of earth and sun and rain will transform the hard yellow kernels into thin green lines across the land. These sacks can turn dark plowland into rippling green oceans of corn, with pale tassels like wind-whipped waves. They can cover the land in the fall with encampments of shocks, spreading over the hills.

All these are concealed in one yellow kernel; all these, and the chance for immortality: for uncountable bushels of future corn lie within this brittle shell, and descendants of this one germ may sprout above a soil of the future, in a world of which we can surmise nothing, and among unborn people whose way of life is yet undreamed.

Corn covers the land of America. It grows upon the great flat fields of the Middle West, a full mile square. It raises its tasseled head on the steep slopes of the Appalachians where, as the mountain folk will tell you, it must be planted by shot-gun from the opposite hill. It follows the sensuous curves of the Piedmont country, and covers the little bottom-land field of the one-horse farmer of the South. No yard is without a row of sweet corn, and the tall stalks throw a zig-zag shadow up the steps of the poor Negro's cabin in the Carolinas.

The land where the great four-row planters plant the acres, and the mechanized corn-pickers devour the ears like gigantic grasshoppers, is peopled with the wondering forms of spectre red men who once had called this home. Phantom Indians who had made the white man this gift of corn, and who danced round the hills they planted by hand, gaze in bewilderment upon this noisy monster that drops the seed into the earth, three to a hill, with as little effort as the flap of a bird's wing in its flight across the sky. But the four-row planter dances its ritual measure and sings its own rough song. And the Illinois farmer, inarticulate though he may be, is aware of the solemnity of this ceremony. Riding the planter across his black ocean of land, he feels the pulse of the engine beneath him, and rejoices in this sense of power. And though he may not pull the planter out of its controlled passage to perform some wild fertility dance, and though his mind is too much occupied in following the furrow made by the marker to be able to sing his songs, yet within his frame lies an awe for these fields and this seed that trickles with such ease through the body of the planter. He accepts bondage to his corn.

The spirits of dispossessed Indians do not feel the same bewilderment as they watch the poor Negro in the South, scratching his earth for his corn's planting. They understand this seedtime, for they know the use of man's hands. As he sits before his cabin, gazing upon the first pale shoots of his patch of corn, it is almost as though the old colored man could reach beyond the actualities of this life and hear strange singing among his stalks of corn. The song he hears is nearer to his own singing than the cadence of the white man's music. He lifts his head to the field, but no one stands there. The singer, perhaps, is at the end of the field, hidden behind moss-hung oaks. But it does not matter to the old Negro that he cannot see him. The song holds the same meaning that he feels within himself. It is a hymn to his corn. The phantom Indian must know that his maize lies in worthy hands.

May is here. The farmer enters his barn for the sacks of seed corn, for it is warm these days, and the leaves upon the oaks are the size of a squirrel's ear. It is the moment for planting, and this descendant of the pioneers, this son of the frontiersmen to whom corn gave the force to open up the West, this twentieth-century farmer loads his truck with the sacks, and drives to the West Eighty. He passes fellow-farmers along the road, their trucks also carrying sacks of seed corn. Upon the fields, as far as the eye can see, tractors draw harrows and discs, that the land may be clodless and fine, ready for the seed.

He reaches the West Eighty.

Into the drums he pours the hybrid seed corn, this outcome of science that creates new wealth for the farmer, this product of hand pollination that, perfect in shape, exact in size, will grow corn to stand tall and straight, corn that can challenge borer and chinch bug. And here, as we watch the Illinois farmer empty his sacks stamped with the seedsman's name, we know there must be confusion in the mind of the phantom Indian who always lurks behind the way of a man with his corn. For back over the years before Europe invaded this continent, fertility and plenty took as their symbol the beauty of the ear of corn, and walls of crude Indian barns, walls of pueblos, walls wherever there was place that they could be hung, were festooned with bunches of corn, suspended by the pulled-back husks. And so, as the white man took from his conquered victim the booty of corn, log cabins over America were decked with these ears, that the pioneer had selected for his seed. Today, as we wander among little farms, we can see upon the wall of the porch a strange iron fork which, even though the farmer be enlightened enough to buy hybrid seed, is used still for the drying of ears. For man lags in his heart where he strides forward in his mind, and the life of the past is stronger to his nature than the science of today. The ears of corn that dry upon the iron prongs, for seed, stand as a needed symbol of fertility and plenty. Man must have visible assurance that all is well. It is not enough to know that somewhere upon the prairie there are seed fields and seedhouses. He must be able to see and handle the ears of his corn.

Over the vast fields of the Middle West move the corn planters, check-rowing the seed. Over the smaller fields of America proud-necked work horses draw the planter behind them, and the rows of corn are planted in straight lines, three to a hill.

Beneath the earth a miracle happens. This gold kernel, flat on its back in the bed of the soil, gives birth to the corn. Out from the kernel, which is food for the germ, spring colorless fibres, feeling their way along this soft earth, for moisture and sustenance. Out from these fibres grow tiny white hairs, searching always for food. Up from the gold shell shoots one strong fibre, seeking the light. And a week or more after this kernel was dropped into the earth, two small leaves pierce the ground, so frail that they seem destined in maturity surely to grow to nothing bigger than a stem of meadow grass. An unspoken shout rises over the land: "The corn is up," and fields are patterned with lines of green, yellow against the sun, delicate like down upon the great brown body of the earth.

Now the corn lies in the hands of a power greater than man. Earth will feed it; rain will make it grow; and the magic of sun will pump the green blood through the leaves. The strengthening little shoot will throw one more leaf to the right and another to the left, and fling down a set of brace roots, to bind it to earth, as lovely in form as the flying buttresses of a Gothic cathedral.

The countryside seems vast now, with the corn so young that it hides nothing of the immensity of these fields. The bewildering pattern of check-planting gives a sense of infinity. Rows converge in a dim blur over the acres, to join more fields of corn beyond our vision. So immense are the fields here in Central Illinois that the bordering trees are minute in size, grey-blue in color, and a tractor pulling a cultivator for the first working of the corn is a speck upon the landscape, like a tiny beetle. There is no need upon these flat fields for the sensuous curves of contour-plowing. All is straight of line, like the green spokes of an opened fan.

Upon this flatness the corn grows, and the world of the Corn Belt thinks and talks corn. Back and forth and across the great fields the four-row cultivators turn the earth; the soil is dark until the sun shall dry it. Damp heat falls like a pall upon the little towns, but the people of this corn civilization, mopping the sweat from their faces, smilingly tell each other that it is good corn weather. For their life is based on corn.

And then, in the heat of July, when the plant is high as the head of a man, it reaches its supreme moment. As the sheath bursts, to release the tassel, we watch the pollination of the corn. With the precise timing of nature, the silks stretch beyond the tips of the young ears. They wait for the dust that means life to them. And the breeze, that stirs the leaves with a harsh sound, silently sifts the pollen from the tassel to the ears.

Over the fields the air is heavy with gold dust. But its purpose has been achieved. The ear swells in the shelter of the waving leaves as rows of kernels form, for the filaments of silk have carried the pollen to each mother cell.

But something new is happening. Great fields of Illinois corn stand dedicated to seed. Today, on the corn land of America, the farmer no longer selects his own open-pollinated corn, with which to plant his fields. Relying upon the men who have evolved the hybrid seed corn, he expects his crop to yield from five to fifteen bushels of extra market corn for each acre that he has grown. So, through the labors of science, strange things take place among the stalks of corn. Trucks arrive at the edge of the fields, loaded with farm girls. This detasseling crew will work through the night, on electrically lighted machines, robbing the corn plants of their tassels. These girls will ride the elevated platforms, between the rows, bending low to the tips of the cornstalks to pull the young tassels before they can shed their pollen. For the ear plants must be kept free from self-pollination. Day after day, through the stifling hours of late July, night after night in the warm darkness of full summer, the detasseling crews pull the potential seed from the ear parents, so that the pollinator rows, one in four, may control the seed corn. It is a world governed by science, with inappropriate-looking paper bags placed over tassel and young ear in the breeding plots, in order that seed may be conserved, and pollination may be done by hand.

Now, through the last few weeks of growth, the corn stands firm over the land. It is so high that it dwarfs the countryside, and the roads are pleached alleys between the fields, with the shadows of leaves and tassels flickering across the sunlit surface in bewildering patterns.

Harvest nears. The first frosts have dried the ear and loosened the husks. Grey comes into the sky, and mud upon the fields. For summer passes and the leaves will soon fall from the trees. And as the sap dries in the plants, the corn fields turn shabby and old. Green leaf pennants that waved so proudly in a summer breeze lean wearily down, ragged and torn, and the tassel hangs its head, against a broken joint of its stalk. Age does not come with beauty to a field of corn.

But within its pale husks rests the ear. Like a bent old woman who sees in her grandchildren a gesture to posterity, the bedraggled corn plant could know it has served its purpose. Tear back these frayed husks, and you will disclose the golden ear, perfect in its beauty.

Now is the time for the harvesting of these ears. The people of America gather their corn, and according to where they live, so does the harvesting vary; for man's method of gathering his crops is a gauge of his place upon the earth, and the character of his culture. They husk the corn in the fields, leaving the stripped stalks standing, an army of shabby ghosts. They harvest the plants with a knife, and the shocks cover the fields like wigwams, waiting to be husked. They gather the ears into the shelter of the farmyard, turning the shucking into a lively ceremony as they sit round the mighty mound of unshucked corn; for man needs his festivals.

Out in the Corn Belt, where the fields cover a mile square, there is no time for such ceremony. And man has had to adjust himself to this, for over the vast fields it is a mechanical corn-picker that harvests the ears. It is like a great silver grasshopper, this pull-type picker; it is like a great scarlet locust, this mounted picker that is placed on top of the tractor. Two rollers squeeze the stalks, and the ear is snapped off, to tumble on to the husking-rollers which remove the husks. Up the elevator goes the ear, and into the wagon, and never is it touched by human hand. Upon the ground behind this devouring monster lie the crushed cornstalks, and the air is flustered with a swirl of husks, like feathers that have been plucked from gigantic fowls. The wagon fills with a gold pool of corn.

But there are farmers even in the Corn Belt who husk still by hand. Over the fields moves the familiar sight of a wagon pulled by a team, straddling one row. There is no sound here of the chug of the tractor, and above the creak of the wagon comes the dry rattle of ears thrown against the bang board. Humans harvest this corn, with husking-pegs strapped to their hands, and so adept are they that they do not need even to turn their heads, to see where to throw the ears.

Everything upon the farm must wait these days until the husking is finished. For men and women are in the fields, tossing ears of corn into wagons. Over in the yard the corn is elevated to the empty cribs, and the muted colors of the farm, under a grey sky, are fired by the gold of the ears.

Down in the Carolinas it is the moment for harvest. The poor Negro and his wife sit on their cabin steps, looking at the shabby corn before them. There is no sun in the sky today, to toss the ragged shadow of the old stalks up the steps. The hogs grunt in the pen; the old couple's one cow stands tethered beyond the little peach tree; and the broken-down crib is almost empty. Out into their corn patch go the two Negroes, for the gathering. There is no ceremony here, and no great monster strips the corn in the blinking of an eye. Two old colored people pick the ears from the withered stalks, and sit at twilight in the corner of their yard, wrenching the precious corn from the husks. For the empty crib must be filled that hogs and cows and chickens may be fed.

The land of America is covered with the aftermath of the corn harvest. Shocks follow the sweep and curve of the hills, the dip of the valleys. Stalks stand like spectres upon the fields. A European, homesick for his own country, thinks of vineyards or little farms. An American, lonely in a foreign land, sees fields of corn. He sees pale rows across the earth in spring, and solid stands of corn in summer; but most of all he sees dim gold shocks upon his homeland in the fall. He sees bright little yellow pools at the foot of these shabby shocks, as solitary figures husk the corn in the first cold of winter. He sees corn-filled cribs, and ears in the shelter of barns. And seeing all these, he will hold fast to them; for he knows these will endure, when empires have fallen and the quiet of the battlefield is broken only by ghosts.

But history turns its pages. With a later dynasty, and the coming of the white man to these shores, in the seedtime of the New World an unknown plant appears. Over the roughly cleared, new-made fields of America a thin spear pierces this alien earth. Patches of vivid green cover the cleared spaces, to thicken and change color in the sun, to stand crisp and gold at the moment of harvest. Among the trees of the uncut forests, slinking Indians must look with bewilderment upon the strange crop. For this was their world of maize, and they sense an intruder.

Pioneer man, in search of new lands, takes with him those things of greatest worth. The European brings the seeds he has inherited from his remote ancestors. In the cramped hold of the *Mayflower* lay sacks of wheat. The pioneer carried the little dried grains that held the power to create, in an alien soil, the landscape of his people. For he must solace himself with the background of his own infancy, and the infancy of his race.

The immigrant today gazes with tenderness upon the ripening grain; his ancestors in the Konavle Valley bent their bodies to the reaping of wheat, and the women, clothed in white, trod the small fields in the pockets of the Dalmatian hillsides, heaping the sickled sheaves. Over this land new citizens find comfort in familiar wheatfields, traitors, at this moment, to America's fields of corn. Before them they see the country of their forebears, and wheat over the flat land of Hungary or on the Steppes of Russia is more vivid to them than the wheat growing in Kansas and Nebraska, until the fields here are peopled with European harvesters, and the Wheat Belt becomes the plains of home, with the songs of Czech and Jugoslav peasants echoing across the fields of grain.

Wheat is a link between the Old World and the New.

It is the Texas Panhandle, late in June. Over miles of flat earth grows the wheat, rippling in a breeze that blows, unhindered by more than a hog fence, from the regions of Canada. Down through the dry belt past the one hundredth meridian it grows, over the semi-arid country of Nebraska and Kansas, Texas and Western Oklahoma. For this is the Wheat Belt of America, the bread basket of the Western World.

The harvesting begins. Three crimson combines approach the six hundred acres of hard red winter wheat. The sun is high in the sky. It has drawn the dew from the wheat, and the combines now can harvest the dried grain. Heat beats upon the harvesters: one man to drive each tractor, one man to tend each combine, one man to oil the machinery; a small crew today can reap and thresh these miles of grain. The twenty-foot combines open up the field with such speed that they seem only this moment to have arrived. Back across the years, men with hand-sickles labored for hours, and evening would find an opened path for the morrow's binder, with sheaves of grain leaning against the hedges of the field.

Hour after blazing hour the combines cross this field, and into the chug of the tractor is threaded the rattle of the threshed grain. A driver stands to his work, his body worn sore by the seat of the tractor. The sound of the monsters is ceaseless in his ears, the dazzle of the sun upon the metal of the machinery blinds his eyes. His horizon is bounded by the great red body beneath him, throbbing in the June heat of Texas. Dusk falls upon this harvest scene, but man is master of the dark, and, their headlights throwing shafts of pale yellow over the standing wheat, the combines work into the night. The only power to stop these harvesters is the dampness of the night air, toughening the stalks. Man is vanquished by heavenly dews.

But though the combines may have been silenced, there is no peace upon these roads. Twenty-four hours a day, seven days a week, now, at the moment of the wheat harvest, lines of trucks flow to the grain elevators of the towns of Texas. And the darkness is split by the beacons of headlights, converging to these towns. Combined upon the fields, the grain is ready for storage; it fills the elevators, or lies piled in great heaps upon the ground outside the buildings. The world of the Texas Panhandle for this short space of time is wheat.

Not far from this scene of twentieth-century mechanization, a different harvesting takes place. In a few minutes of flight by airplane we can roll back the centuries, for across the state border, in New Mexico, Indians reap their little wheat fields with a hand sickle, as man has harvested his grain from the beginning of time. Through the ages man has bent his back to the reaping of his grain, and earth has been patterned with sheaves. Combines hold massive beauty, but they have deprived us of that heritage, the evidence of harvesting; for these monsters that behead and thresh the wheat leave the stalk standing, as though the fields had been pillaged rather than harvested; and there are no sheaves. Man who has created the machine will need more years in which to forget, in his instincts, the ways of his forebears. We live in a civilization between two eras, for our minds accept the machine even while our emotions turn back to the past. Today, as we watch the powerful combines harvesting the wheat of the Texas Panhandle, our brains tell us that this is the only way to feed a hungry world. But the belief in mechanized harvesting is tempered by a longing for the beauty of sheaves, and we toss our minds back to days when bodies stooped over these sheaves, like figures in a dance, and arms curved in a wide circle, to enfold the reaped grain. And suddenly we listen for the whirr of the binder.

If we would hear this sound we have only to walk in the wheat lands of the Eastern States, where the fields are frequently too small for the combine. A sense of the universality of harvest fills us, as we watch tractor-drawn binders pass along the highroads to the fields. We see before us great elms in an English lane, and the dripping branches brush the heads of the horses that draw the binders. Seeing this, we know once more that the strongest bond among all peoples is labor upon the earth; for upon the earth all men are the same. The English harvester halts his team, as a pheasant is flushed from the field of grain; the little ones may still be there, less able to escape the blades of the binder. His American brother brakes his tractor as a quail rises from the wheat before him; her little ones, too, may be hiding among the roots of the wheat. Man's humanity to small creatures knows no frontier. The squawk of the pheasant or the call of the bob-white is strong enough to be heard above the binder's whirr.

The wheat has been cut. The harvesters have made their way in the path of the binder, gathering the sheaves, and the fields are patterned with shocks. In their varying forms they follow the contours of the land, creeping over the heaving fields, to be lost to sight in the valleys. They are illumined in the hours of the morning, and stand dark against the sun as it lowers in the sky. Shadowless at noon, they cast long lilac arms across the stubble at evening, stroking the shocks in adjoining rows. If they could only stay upon the fields through the weeks of the fall, we feel, as we gaze with contentment upon these tokens of abundance. . . . But we hear the creak of wheels along the lane, and the sound of a man calling to his team. Into the field comes the wagon, for the hauling of the sheaves. We look at nightfall upon deserted stubble. Loneliness lies over the land.

In the Texas Panhandle, where the wheat has been combined, there is no such sense of desolation. The plow has turned the earth, and a summer rain has fallen. The fields have changed from pale gold to a brilliant green. For the combine had shattered the ripe grain, and in the warm wet soil it has germinated and sprouted, covering the fields with a garment of volunteer wheat. And now these pastures are dotted with sheep. From the desert mesa country they were shipped east, toward the meat markets of Chicago; breaking their freight for six months, they are fattened on the sweet young wheat of the Texas Panhandle. From Colorado they come, for the meat markets of Kansas City, and day by day the animals gain weight on this lush offering of the combines.

But the wheat in the East is stacked in the farmyard, awaiting the threshing machine. One morning, as we walk through the fields near the farm, we shall hear the chug of the tractor and the rattle of the thresher, and see a long chute spitting the chaff upon a great buff pile. Our throats will choke from the dust of the wheat, and above the noise of the machines will rise the laughter and shouts of the workers. For there is a feeling of carnival today, as the custom thresher reaches the farm. This is one of the year's festivals, like corn-husking and butchering, which hold the power to gather men together in their labor.

We can trace the history of immigration to America through the types of grain that the settlers brought with them, but a greater span of history is mirrored in the milling of wheat flour. This grain, whose exact origin is unknown, and whose ancestry is dimmed by antiquity, must have had its beginning in the cradle of the human race. For with man's cry for food came wheat; wheat to be ground into flour; flour to be made into bread. In the great roller-mills of Minneapolis, science has reached perfection. But it is the same grain that grew in Babylon and Nineveh, and the same grain that fed the people of China many thousands of years before Christ.

If we could turn back from the rolling-mills of Minneapolis and the combined harvests of Kansas and of the Dakotas, to the world of our ancestors, a few centuries ago, we would understand what civilization has meant in terms of wheat. We would harvest with a sickle, over the tiny fields, the grain we had sown by hand. We would see cradles of stone beating the grain from the stalk, even as today, in little islands of the Mediterranean, blindfolded ponies draw the stone cradle over the circular threshing floor.

But it is in the actual milling of the wheat that we watch history. As primeval man reduced his grain to flour through thousands of years by means of a handstone, so, in England, stone remained supreme, and the bread of our forefathers was made of flour ground in a quern. And as the common man seems to have had always to fight for access to food, so, across many centuries, he battled against lord and abbot for permission to own his quern and grind his own flour.

Over the harvested fields of past centuries, this common man gleaned his pitiable sheaves of grain, that he might be fed. Today, with the world's wealth and its harvests, there should be no longer need for the gleaner. Civilization should mean food for all men to eat. For the history of a civilization is not the history of its battles. It is the picture of man's struggle against starvation. The shape of a nation has been molded by the way it was able to nourish itself, for human progress means the increasing ease with which man finds his bread.

Before war enveloped them, trampling the fields of grain, there was a people in Europe that lived mainly upon wheat. It was the peasantry of that continent, working among goats and vines. The smell of long crusty loaves rose in the air from village bakehouses. There was laughter in that world, and delight, among the Sicilian peasants with their gaily painted donkey carts, and the goatherds on the mountain sides of Greece. Their laughter stemmed from the eating of good food; their delight came from labor upon the earth. Strangely enough, a certain Illinois farmer, with the physical bearing and moral values of Abraham Lincoln, might have understood these colorful people better than most; that farmer who told us once, as he talked of the contentment in the lives of his hired men, that all the world really needed, for happiness and peace, was a feeling of confidence, and good feed. Perhaps, across the continents, this farmer, so alien in his blood to these Latin peasants, might act best as a merger of race and culture. For he knows what happens to a man when he grows his food upon his own land, and can delight in the harvest of his labors.

## 4: *Meat*

"Tomorrow I've got to take that bunch of steers by truck to Chicago," says a stock farmer in the Middle West, as he points to a clump of penned cattle in the yard on the top of the rise.

The slopes below us are sprinkled with a herd of Black Angus; cows with their calves, heifers and steers. It is an idyllic picture, with the thick grass high around the legs of the creatures, and a stream winding among the roots of the trees. The farmer turns to this scene, and there is gentleness in his face.

"You won't find me buying feeders if I can help it," he says. "Why, a man loses half his pleasure that way. What would I do if I couldn't watch my cows with their calves? One of the prettiest things in farming is to go out to your pastures at sundown and see the mothers go off to some corner of the field for the night, each with her own little calf. There's a tenderness there that does a man good. . . . Not that it doesn't cut both ways; for when I've known my animals from the start, and planned them before they were born, and can tell one from another across these bottom-land timber pastures, I feel a traitor when I take them to the packers."

To our right, where the slopes rise to the level of the road, a little bull stands tethered. It is heavy of shoulder and soft of eye. The farmer walks towards the animal and pauses, fondling its head.

"When I think of the calves I shall get from him," he says, "it would be just foolishness to let some one else have all the fun of the breeding. If I can go to my grave knowing I have added a good herd of pure-bred Angus cattle to this world's riches, I shall feel I have done something to justify my life on earth. To have bred a fine animal, or raised a good crop, seems to me worth living for."

The pastures grow dim before our eyes; for we seek within our minds something similar that was said generations ago by a man who had never been to America. The words of Dean Swift come clear to us now, voicing the philosophy of this stock farmer among his cattle:

"Whoever would make two ears of corn, or two blades of grass, to grow upon a spot of ground where only one grew before, would deserve better of mankind and do more essential service to his country than the whole race of politicians put together."

The farmer leaves his little bull, and walks towards the penned cattle in the yard.

"Yes," he says, "I've got to take that bunch of steers to Chicago tomorrow. My father drove cattle to Chicago in 1854, and his family has had cattle on the Chicago market every year since then—reds and roans and shorthorns."

Seeing him next morning with his Angus steers loaded in the truck, starting at dawn for Chicago, we think of the days before the automobile, when the father of this man drove cattle to the yards. And the drama of the trade excites us, with the romance of the range and the legend of cowpuncher and cattle-brand. We feel the lure of the Chisholm Trail, and hear in our ears the crooning song of the cowboy, at the bedding-down of the cattle. We see chuckwagon and lariat, and share the riotous gaiety of the cattle towns of the past: Ellsworth and Cheyenne, Dodge City and Denver. We watch the pageantry of the roundup and search the prairie for mavericks. Into our nostrils comes the sizzling smell of the branding iron, and the scent of coffee for the cowhands' breakfast. And we remember that it was only the year before last that, driving across the West, we saw before us the magic words: "Open Range," and bumped with our car at night into the ghostlike forms of straying cattle.

Romance dwells always upon pleasing things. Thinking back to the history of the Western cattleman, we know that there was another side to this entirely American drama. . . . The cowboys upon the ocean of grass watch the Texas sky. There is a restlessness among the cattle, which betokens a storm. Into the ominous stillness of the evening rise the mournful tones of cowboy songs. Over and over they croon these songs, pleading with their cattle to lie down. The tenseness in the air leaps like electricity from cowhand to cowhand, adding keenness to the ears that listen for the night sounds of the herd. They wait for the cattle to bed down. They listen for them to hold their breath for a while before they blow off. Over from the west comes a rumble of thunder. Five thousand longhorn steers bed down to their herdsmen's wailing lullaby.

Something stirs. A steer rises to its feet. Soothingly the men sing, to calm the creature. But the thunder has more force in its ears than the singing of men.

And then it happens. This most terrific of things, a stampede, starts. It starts with a unanimous suddenness that scares. It is like a current switched on, firing five thousand animals. Generating their own electricity of terror, they are like the storm that rages now in the sky, brought down to the earth of the prairie. Lightning splits the heavens, catching the horns of the animals, until it is like the center of exploding fireworks. Thunder roars in the sky, but there is no sound from the stampeding herd. They are running straight, without clicking of horns. A great heat is thrown out by the fevered animals, merging into the heat of the storm from the sky. It blends with the sickening odor of hoofs and horns. . . . But they have stopped stampeding now, and must be milling around in the storm, for a great bawling arises, as loud as the thunder. . . . What daylight will disclose is a dream which lies in the nightmared minds of the cowhands.

These are ghosts we have conjured up, for the longhorn has vanished, and the cattle herds of the West are confined today within fences of barbed wire. But a civilization needs to respect its ghosts, because they are the image of its ancestry. And the ghosts of America sit the saddle at a stampede; they do not walk in the clinking armor of their European brothers. This stampede is the heritage of America's cattle trade.

But we listen now to more recent happenings. Into this ghost tale come folk legends of the long trail-drives of old; and great rivers are crossed; and cattle are drowned as they mill in the water. Story follows story, for Jake Shank is full of liquor, and the evening is young. He leans against the counter of a store near Albuquerque, a typical figure of an old cowhand. His long plumed moustache is grizzled, his skin so dry that it is as though the bones of his face were covered with stretched rawhide. But there is grace in his old body, and the sight of his compact hips and strong-muscled thighs still hold the power to delight any woman. As he stands here, his wide black cowboy hat shading his face, the screwed-up eyes seem always to be looking far, searching the brush country for cattle. There is something of pathos in this figure that has outlived its era; but he is not aware of it himself, for his world is within him, strong and invulnerable. He sits his saddle endlessly in his day-dreams, choking with dust kicked up by the cattle along the Chisholm Trail. With those strong arms he swings the lariat and ropes a yearling bull. He halts his horse and dismounts, to tie the bull's feet together with his piggin' string. Louder in his ears than the sound of automobiles on their way to Albuquerque is the imagined bawl of the big bull calf as he dabs the heated branding iron on the left hip. In his nostrils he can smell singeing hair and burnt flesh. There were some who said you couldn't do all this alone, but must have two flankers to pounce on the animal and hold it down, and a cutter to castrate and ear-mark it; but he, Jake Shank, can brand a range bull himself. He leans forward now, looking suspiciously round at the other people in the store before he speaks. And then in a whisper he tells his secret, for it is a secret he has kept to himself all the days of his life: never can he get used to the hot smell of blood and burning flesh, and he, Jake Shank, the champion cowhand, will turn always from the branded bull, the running iron pink still in his hand, sick at the stomach.

Jake Shank does not know how much his world of cattle has changed. If he did, he would pity the cowhands crouched in the caboose of the freight train that carries their beef herds to the stockyards, as he thought of days of sun under the Texas sky, when the cowhands sat in the saddle, driving their cattle to market. He would shake his head at the chutes and stanchions, where the animals are branded standing, for there is something in the muscles of the cowpuncher's shoulder that frets for the swing of the lariat; and were he to hear cowboy songs around the camp fire of a dude ranch, with no bedded-down cattle in sight, those screwed-up eyes would sadden at such abuse of tradition.

Care for him tenderly, this aged cowhand; for he is America's history, with the cattle-bones turned up by the plowshare in Texas, and the Indian arrowheads hidden in the furrows of a field in Ohio. But do not pity him; for he has lived abundantly, upon dust and sun and kicks and tumbles and the smell of the breath of his cows.

Though the days of the trail-drive have gone, there are some things still that would delight the heart of Jake Shank. Out in the West, where the land stretches farther than the eye can see, are ranches so vast that even their owners can get lost upon them, and the cowboys have as their sole companions the wildcat and deer. This world of great spaces is a world where men are known by the names of their cattle brands, knights of an aristocracy of the wild, with a heraldry as guarded as that of any ancient lineage. For cattle "rustlers" abound, who would alter the brands and steal the cows, and what, to us, has the beauty of ritual, is rooted in necessity. Rail, bar, mashed circle, box, broken arrow; thus are some of the units of branding, burnt upon the hides of the beef herds, that no thief may change them.

Over the prairie, at the time of the roundup, the night guards patrol the sleeping herds; and the songs they sing, to lull the cows to sleep, are the songs of their cowboy forebears, from trail-driving days. For man evolves the technique best suited to his trade, and the craftsmanship of the roundup today, from Canada to Texas, has been molded by cattlemen over the generations from the sixteenth century, when calves were first brought to the New World. Across the earth man sings to his animals, gathering them to rest. "Coo—sheepie. Coo—sheepie" calls the shepherd in the mountains of Virginia, and his flock flows to him, like tributaries towards a river. On the mountains in Greece, shepherds played reed pipes to the herds that cropped among the gods; and the little Breton peasant girl assembles her medley flock of sheep and goats, cows and geese, at sundown with a plaintive song. The cowboy's lullaby is in direct tradition.

They have roped and branded the new calves, and the beef cattle are herded for the market. But their destiny is not yet stockyard or packing-house. They must go East, to the Corn Belt, these short yearlings and long yearlings, these white-faced Hereford steers. Out to the harvested corn land of the Middle West the freight trains carry them, to glean over the stalk fields of the fall. And the flat fields of Iowa and Illinois, Nebraska and Missouri, are dotted with these white-faced feeders, until the cold of winter withdraws them to the barns for a diet of grain and hay. Through the long months of winter, rich in color against a world whitened with snow, their breath like clouds of smoke in the biting air, throughout the long winter of the Middle West, the beef cattle grow slowly heavier, ever nearer their doom. Horizon bounded by farmyard and feed-bunk, the memory of calfhood in Texas dims, and sadness comes into the gentle eyes. Their lives are circumscribed by eating and sleeping, and even the delight of their maleness has been cut from them. Rows upon rows of Hereford steers, they stand in their pens and their stalls, awaiting the packer.

For the cities must be fed.

Warmth comes at last to the air. Spring, which is with us in the South almost at the turn of the year, tarries here, in the Middle West. Winds sweeping down from the Arctic Circle kept the cattle within shelter of the farmyards, but today a soft warmth blows up from the South, over the plains of Texas, from the Gulf of Mexico. Growth thickens the clover fields and the pasture lands of blue grass. And as the warm wind brings spring to the North, thawing the prairies, an unrest stirs within the penned cattle, and a craving for light. Into the sun they must go, that our meat may be richer; out to the meadows of clover, tasting so lush. And the clover that is grown so that nitrogen may be restored to the soil, thickens and blooms, through spring and summer, to be clipped by the soft mouths of baby beeves.

The Corn-Belt farmer walks among his cattle in the cool of the evening. Before him he sees a good bunch of white-faced yearlings, fattening on rich pasture over the months of summer. Soon, when the corn is husked in the fall, these steers will go to the feed lot for a full finish of corn. They will put on weight and fat, to be ready in early spring for the stockyards of Chicago. Their destiny is planned, without possible deviation. No deliverer will appear, like a prince in a fairy story, to save them, by reason of their beauty. For their purpose in life is to be turned into meat, and Texas ranges, Illinois stalk fields, hay, clover and feed-bunks of new corn are prescribed solely for this end. Gentleness of eyes, or grace of curling locks, has no significance. They are beeves, to be assessed by weight.

Through the sweet air of a spring night, on their second journey home, a bunch of well fattened Herefords shakes in the rattling cars of a freight train, bawling in discomfort. Morning finds them being driven from the cars down the lanes between the pens in the stockyards of Chicago. They pass steers from the land of their birth; half fat and grass-fed cattle, off the ranges of Texas. The stockyards are filling with cattle, as salesmen on horseback lean down, to open the pens. The spring air is thick with the smell of frightened animals, loud with their bawling. Along the walks over the tops of the pens stroll the packer buyers, looking down upon the cattle. Here, under the open sky of Chicago, in rain or in sunshine, animals are sold on the hoof in their thousands, by deals between packer buyers and commission men, to be sent to the great meat packing houses where they are butchered with the speed of an assembly line.

For man must have his hamburger and steak.

He must also have his bacon and pork and sausage. Norman Starr knows this, as he walks his farm among three thousand hogs. The elderly gruff man lives and dreams hogs. Speak to him and he will answer you: "The only thing I know is hogs, but if you'd like me to talk about my hogs, I will." He will, and he does, over the hours; and anyone who would say that the raising of hogs lacks emotional drama should listen to Norman Starr.

He stops to look at the yard, filled with his belted Hampshires. There is beauty here, with the contrast of the black bodies against the yellow pools of corn in the feed-bunks, the light color of the belts breaking the pattern like a jig-saw puzzle. And the hog man must be feeling this beauty, for he turns to us with softness in his face as he says:

"Why did I start in on belted Hampshires? When I saw that belted Hampshire, I tell you, I fell in love with him—or perhaps you might say, I fell for him. Of course, you could also say that he's a pretty thrifty hog, and a good grazer on clover and alfalfa. And he's wide awake and active, and gets around well in the feed lot."

We wander to the hog-wallow. It is a strange primordial world that we see, heaped with rounded humps, in a vast pond of mud. And then, as the humps glisten in the sun, they become the bodies of numberless hogs, covered with this mud.

"But don't you go and think I let them do that when they're young," says the hog man. "We've other ideas these days, with our County Swine Sanitation. You should just see us around farrowing. You've noticed all those little houses over there? They're my individual hog houses, with moveable floors, for the sows and their litters. I don't mean to have my little ones down with baby pig disease. It breaks your heart, to watch them dry up and crawl off and squeak and spread out and die, and nothing that you can do about it. To think of all the diseases lying in wait for my pigs, it does something to you, I can tell you. There's the round worm, and erysipelis and Brucellosis, and necrotic enteritis and——"

But he breaks off to smile at a pig. There is no doubt of it, he loves his hogs. Tenderness glows in his face when he describes the farrowing of his sows, and the great clumsy hands make the movements of scrubbing the animals' bodies, and washing the udders. There is sadness in his voice now, as he tells of the spring farrowing this year.

"It was fifteen below zero when they started to come," he says. "It was that awful cold spell, in the middle of March. We did all we could, to save them; but they died, so many of them, from the cold. It nigh breaks my heart even now, when I tell you folks about it, to think of the little bodies lying around, stiff and dead, and they so beautiful and all. . . . But there's the fall litters will be coming the last of August. I've got two-hundred-some mothers due. It's a lot of trouble, the labor of babying the pigs, but it's a pretty sight, to see them running around in the clover, when you've hauled them out to a rotated field—all those little things not more than a week old. There's some that go crazy over lambs, but give me a baby pig any day—and a belted Hampshire at that. There's no beating a belted Hampshire, I tell you. When they go on the hooks they're not as lardy, and those packers in Chicago like the looks of them, they do."

It is clear that Norman Starr has only added the last sentences that he may not seem sentimental. There is a forced severity in his face as he swings his talk round to the destiny of his hogs. These creatures that he loves so well will be fattened and sent in double-deck freight cars to the stockyards of Chicago, to be butchered at the rate of about five hundred and fifty an hour, with the efficiency of the packing house of today.

Thinking of this, our minds turn to other butcherings. We see a family of Negroes in the South, at the ceremonial killing of their only hog. And the tenderness that Norman Starr has for three thousand swine is concentrated into their feeling for one hog, suspended from the hickory tree in the yard, behind the cabin. Gently they wipe the carcass of their hog, as though it were a sacrificial rite, and all ugliness is drenched clean from the ceremony by the reverence in their gestures. Over America, with the first sharp snap of frost, smoke rises into the cold air from fires that heat the water in scalding barrels, and carcasses of hogs hang suspended along poles, against a background of Appalachian mountain, a background of tobacco fields in the Carolinas, a background of Middle West prairie. For the meathouses of America, in the farms across the land, must be stocked with sides of bacon, sausage, and hams.

Cattle and sheep and hogs roam the pasturelands of America; from

Wyoming and the Dakotas, Texas and the Middle West, from prairie and feed lot, they are drawn to the great meat markets of the country, to become our food. But behind the cattle trade stand figures like the farmer who loves to watch his cows at nightfall, gathering their calves to the shelter of their sides, and Norman Starr who would sell his soul to his hogs; and in Illinois there lives a stock man who, once a year, brings a party of cowhands East, from Texas, that they may see, for the first time in their lives, his finished cattle. For he knows, this Illinois farmer, that man needs to be proud of his animals, and that something good will happen within the hearts of these Texas cowboys when they see those white-faced beeves, that they knew only as wobbly-legged calves, new-born upon the range, grown now into well-fattened cattle. The diary of Samuel Keller's grandfather, upon an Illinois farm in the year 1837, states: "To butchering a sheep— 12 cents. To Auttering a bull—62 cents." We have travelled a long way since this, and there have been many changes; but one thing endures, and that is man's love for the animals he tends.

## 5: TRUCK FARMS

As the freight trains move over the land, their cars carry lettuce and beans from the great flat fields of the Coastal Plain. Up from the warm earth of Florida to the ice-bound cities of the Atlantic Seaboard come the loads of green vegetables. The trains rumble through the night, and people asleep in houses near the tracks turn in their beds as the noisy cars roll by. Within these refrigerated cars are salads raised upon the truck farms of the South, where the sun is already powerful for growth.

Winter moves forward, and the earth grows warmer under the strengthening sun. The territory of harvest shifts northward now, at an average of fifteen miles a day, as the earth turns on its axis, and freight trains bring to the snow-covered cities of America vegetables from Alabama and Georgia. Soon, as the days of spring lengthen and the sun grows yet stronger, this harvest will move steadily up to the Tidewater of Virginia and the Carolinas, and pickers will stoop to gather lettuce and beans upon the flat truck farms near Wilmington. Negro gangs will harvest broccoli and turnip greens, radishes and spinach, and the land will echo with their singing. Trucks will rush these vegetables from the fields to the waiting freight trains, for the cities of America.

As summer draws near, the pickers will bend over the shadeless fields of Maryland's Eastern Shore, then over the great flat trucking country of New Jersey, then over the potato fields of Maine and the earth of Canada. And the freight trains will flow South, now, to the cities; for the sun is high in the sky and the earth is at its moment of greatest heat.

But not all these vegetables find their way into the cars of freight trains. Trucks speed along the highways of America. They are filled with wooden

boxes, and these boxes contain canned goods. Within these cans are tomatoes, beans, sweet corn, and peas. The trucks are driven to the cities from the canneries of the East, where the harvests of the stretching fields were gathered and processed, as food for the months when the earth is cold and no green thing grows.

North and south, east and west, move the harvests of the earth. And behind these tomatoes and beans, these carrots and peas—as behind all harvests—lie plowing and discing, seeding and cultivation, the anxious searching of the skies for rain, and the fear of flood. All these are within the cans, and the bending and picking in the heat of summer, the dazzle of the sun upon shadeless fields and the weight of filled baskets.

Here in this trucking country of the Eastern Shore, the whole land is an enormous plain, flat to the horizon where fields meet sky. That sky today is smooth as blue porcelain, and clear without flaw; its smoothness offsets the patterning of this earth. For the earth is like a great loom upon which are woven many designs. The shuttle that shoots the threat of woof across the warp is a harrow, drawn by a scarlet tractor.

With the mutability of life, which changes even our standard for beauty, there seems to be magic today in the squat shape of this tractor, moving over the land. Eyes that used to turn with nostalgia to the slow rhythm of a horse-drawn plow, and had watched with distress the march of this red monster, seeing no grace in its functioning shape, suddenly are aware of beauty. And then we discover what has happened within us. The tractor is one of the few forms of mechanization today which are entirely benign. A world at war, geared for killing, with tanks that cross the earth to destroy everything before them, shifts our values; looking now at the tractor, as it draws the harrow across the fields for the growing of food, we know that we watch construction rather than destruction; and we see in this powerful shape something to be praised. For it weaves into the fabric of the world the pattern of a well-fed people.

This is a landscape of furrows. Great bare fields are plowed and rolled, disced and harrowed, and these furrows turn the earth into a brown corduroy coat, laid flat upon the surface of the land. But it is a coat of varied weaves: narrow, coarse, wide, criss-cross weaves like the chocolate tartan of an imaginary Highland clan, complicated weaves as though a woman sitting at her loom had decided to try her skill upon a new design. They stretch across this flat land till they touch the horizon, sharp-edged where the shuttle of the tractor but lately moved over the earth, soft where they were woven a few days back and rain has beaten them down.

But though these lines are unchanging across the surface of this land, the color of the earth varies like a flecked tweed. It varies within the bounds of one field, until, as we look upon the mottled scene before us, we watch for the shadow of a cloud to pass, and, seeing no movement across the earth, look upwards with amazement to a cloudless sky. Only as we see cows grazing upon some pastureland do we realize what these fields are like; they have the brindled colors of a little Jersey cow. Blueish brown, blackish brown, warm rust-red, grey, buff: all these shades merge upon the tapestry spread before us.

Into this tapestry are sewn threads of green. The beans have pushed their heads above the earth, thin of thread as we look along rows which converge at the end of the fields, thick and richer in color as we pass beans that have been sown across a field, the way of the road. Later, as the plants grow larger, the patterning of this earth will be lost in exuberant growth, and the landscape will change from furrowed browns to a mantle of shapeless green. Just now, in the spring of the year, we can see the structure of this growth, and look with excited hope upon lines of young beans and potatoes, and the limp, fretted leaves of newly set tomato plants. It is a time of surmise, more stimulating than achievement, and looking upon these bare fields, covered with hieroglyphs of harrow lines, which curve like the gyrations of an ice-skater where the tractor turns at the headland, we can plant them in imagination with cucumber or cantaloupe, sweet potato or corn. The destiny of this land already is planned; but it is hidden from our eyes.

Here, in this trucking country, the deep brown of unplanted earth and the green of the beans are emphasized by the tender color of great fields of peas. It is a gentle green, mixed with the milky quality of the peas in flower. They spread over the land, and before long the quiet of the countryside will be broken by the shouts of pickers, gathering peas for the waiting maws of the canneries.

But the land just now is not invaded by pickers. It is the season of growth, when man stands back and leaves it to sun and rain to swell his beans and tomatoes. For the year is still young.

Over the big truck farms today the tractor traces its hieroglyphs, preparing the land; upon the small farms the farmer and his wife sow cucumbers. Two figures cross the field, dramatically erect upon the flatness of this country. They carry the whole feeling of man's work upon the earth, for they labor with their hands, unaided by machinery. Slowly they cross the field, the shambling farmer walking in one row, his sunbonneted wife in the next. He walks ahead of the woman, and there is a sense of drag, as though she were trying with difficulty to keep pace with her man. With a dance-like movement they sow cucumber seeds, four feet of distance between the hills, five feet between the rows. Two strides of the farmer's legs, and he bends low with the gesture of a figure in a square dance. As he bends, he swings his right arm around to pick out from the old tin saucepan he carries the oval, fawn-colored seeds of the cucumber. Across the length of the field, back the length of the field, until the surface of their earth is planted, time after time, with regular spacing and identical movements, the man and his wife place the seed into the earth and, rising erect, give a dance-like twist with the left foot, to smooth the earth upon each hill. By early July, seven acres of cucumbers will be harvested.

Over the surface of the Eastern Shore such figures plant their fields. They plant them with the variations found always where man retains his individuality. And so we see that one man with his wife will swing the left foot across the hill of earth, to smooth its surface, and another will give a little petulant stamping gesture with the right foot, while others will creep along the earth upon their knees and smooth it with their hands. Two Negroes move across that field to the right, past the clump of farm buildings. As they stand out black against the dark earth, there is majesty in the slow dignity of their sowing; the man makes a hole in the earth with a stick, and his woman behind him, gay in a pink cotton dress and a primrose coat, inserts a small plant into each hole he has made. As she moves, her gestures hold a sense of willing servitude to her man, to their earth, and to the tiny plant she places into that earth.

Separated from this scene of primitive cultivation only by a clump of trees, though it is removed in spirit by an epoch of civilization, a tractor moves over a field with speed hardly less slow than the walk of the Negro with his stick. Behind the tractor sit two colored men. They lift sweet potato plants from a tray and drop them down through the mechanical planter;

upon the field this tractor leaves two threads of standing green plants, like the trail of some prehistoric insect of the time of the dynosaur.

Just now, at the spring of the year, this trucking country is a world of brown and green. We feel the need for color, knowing we must wait for the scarlet tomato harvest to flame over the land. The only bright color we see now is the clothing of the women who plant the fields. Upon these great oceans of earth they are gay little ships, the wind in their skirts looking like sails that billow in the breeze.

Farther north the asparagus harvest has started. At first sight the field looks shabby and destitute, and we might wonder what it was that held the power to gather this crowd of brightly dressed, noisy Negroes. Over at the far end of the field, by the railroad track, stands a truck, waiting to be loaded. These colored people are asparagus cutters. One fat woman comes towards us, a bushel basket rested upon her shoulder, her arms held high, and there is a lumbering grace in her walk as she balances the weight of her burden. She sings as she moves, a song as wild and colorful as the clothing she wears. And so exuberant is she that it seems in keeping that she should cover her fat dark body with pale green-blue trousers and a crimson-and-white striped overall, and wear upon her head, pushed sideways by the bushel basket of cut asparagus, an enormous straw hat.

They have the abandon of all colored harvesters. Fat women burst from torn blue overalls, in complete unconcern over the bulging of their dark flesh. Slim young girls move with enviable beauty, rhythmic gestures in the curve of their arms and the bend of their torsos. Each movement is unconsidered, determined by necessity. This beauty before us, upon a muddy field on the Eastern Shore, is to be found always, in any harvesting. Ruth gleaning in the barley fields of Bethlehem, German peasants gathering grapes in the valley of the Rhine, Frenchmen cutting artichokes in the shade of little peach trees upon the Mediterranean, American farm women picking beans for canning: race and creed is of no concern against the rhythm of bodies that bend over crops they have raised, for in the curve of these arms that harvest we see all loveliness.

Soon, as the weeks roll by and harvest succeeds harvest, color will blaze upon the fields. It will blaze in the tiny fruit of the strawberry, and the gatherers will cross the fields as in humility, upon their knees, shuffling along the rows in their search for the hidden fruit. And while city-bound people crave escape from heat, into the streets of the cities comes the nostalgic cry of the strawberry-seller, shouting his wares. Upon the truck farms of the country, fruit-filled baskets burn crimson, an offering to the people of cities who are deprived of fields and trees and sky.

June comes; the heavens are colorless with heat and the earth misty with heat. Upon the flat land of the Eastern Shore there is no shade. Neither is there respite for the harvesters. Crops mature, and will not wait for a heat wave to break. It is the moment for string beans. Early in the morning, with the sun still low in the sky, the heat lingers from yesterday. The picking crew plunders the bean fields. Throughout the ruthless heat of the day they ravage the rows, till eyes are so dazed that they can no longer tell bean from stalk and must leave it to the touch of fingers to determine the difference. Evening brings no relief from this heat, which seems to pull the pickers into the very earth they tread, and the sun sinks behind the fields, a ball of angry red in a stifling sky.

Over the land trucks loaded with filled bushel baskets have moved to the waiting canneries. Darkness alone has stopped them. But night brings no rest to the pickers. They see before their eyes, through the long hot hours, rows of beans upon a field, stretching to eternity. They search in their sleep among the leaves of the bushes, endlessly picking beans, and, as the hands of the Illinois farmer, in sleep and perhaps even in death, feel always within their grasp the form of an ear of corn, so these tired creatures pick through the night and fill, in their dreams, uncountable bushel baskets of string beans. Tomorrow, with its heat and stooping and gathering, will be but an extension of tonight.

One more moon stands as a sickle in the sky, fulls, and wanes, and the harvest changes. The fields are covered now with tomatoes. July is here, and the heat is even more fierce than it was in June. The eyes that search and the hands that pick would give much today if their harvest were the cool green of the bean. The flaming red of the tomato is like innumerable setting suns.

"You bend down and see stars flying," says one of the oldest of the gatherers. "It's awful hard work, picking tomatoes."

Yes, it's awful hard work.

Upon the truck farms of Florida, in the Imperial Valley in California, on the tidewater land of Virginia and Carolinas, along the coast of New

Jersey, the earth is patterned with long rows of vegetables, for cities far away. Lettuce and peas, broccoli and onions, potatoes from Idaho and Maine, white celery from the black earth of Michigan: from all quarters of this country they come, for the people who have no earth upon which to grow their food. And the shelves of the markets and food stores are crowded with cans, which have been filled with these harvests.

Out by the edge of the fields of truck farming stand the canneries, greedy for the fruits of the earth. Within these canneries is drama and rhythm. It is like a great ballet, the inspection belts and elevated conveyors tying figure to figure, the décor colored with the bright green of swirling spinach or the scarlet of massed tomatoes. Today, as we enter the building, they are canning asparagus. Trucks have brought the baskets we saw being harvested by singing Negroes. Out on the land all now is quiet, for the "field run" has been cut today. "His asparagus got away with him because of the rain," sighs the man with the sampling box, as he sees how many heads have started to flower.

The bright colors of the harvesters are wafted in here, to the artificial light of the cannery. The girls who sit at the sorting conveyor are like rows of flowers in a garden. Dark hands lay the asparagus in rows, and the saw cuts off the tips. Along canals of belting flows the stream of tips, and the girls watch and sort. Up an elevated conveyor the stream is carried, to plunge into a rotary cold-water washer. Into a blancher it must be flung, to emerge limp and bright green.

In the next building a quiet old Negro man sits with a big rake, heaping great piles of spinach into metal pans. He is the one silent thing in this canning factory, where belts and conveyors make their own assertive music. Sitting here in the midst of this machinery, he has the look of unreality that we see in the human who appears in a puppet show. There is more drama in the co-ordination of belting and machine than in the people who tend them. Can it possibly be that this can-conveyor, this creature that belongs surely to a macabre ballet, this thing of metal that gives the twist of a sentient individual as it brings the cans down to the circular filler, has no power for thinking? Each can is dropped into its own hole, to be heaped with garish green spinach, to pass and twist and twirl like a ballerina. Along the conveyors move the shining cans, smoothly, yet with little hesitating jerks, as though they paused to catch their breath in this grand chain of the dance;

in and out, set to partners, as the brine water is poured into them: what does it matter that they are shiny tin cans, filled with spinach, with asparagus tips, with scarlet tomatoes? They have the rhythm of a dance, and an exactitude that might well be the envy of a mortal. Over there, at the sealing machine, see the gentle way each top is pressed upon each can; the movement has emotion in it, as though a mother were placing a cap upon the head of her little child.

But the cans now are sealed, and never shall we see the red of tomato or the green of spinach, until we bring back to our home a brightly labelled can from the shelf in a food store.

Somewhere on the fighting fronts of this war, men are eating canned beans that were grown upon the fields of Maryland. They are eating dehydrated carrots and sweet potatoes; but they did not see them when they were packed, brittle flakes with the opalescent quality of the seed of the sweet corn. They did not see them as they lay, golden in color, upon trays that slid into the counterflow tunnel, where the giant fans dried all moisture from them. Over the seas they go, the great tin cans, and these brittle carrots and sweet potatoes are restored to softness by alien water that did not nourish their roots beneath native earth. And yet, as a soldier eats his dinner somewhere in the war zone, it just may be that the water which cooked his dehydrated food has been drawn up by the sun from a stream in Virginia and blown east across the Atlantic in a cloud, to fall as rain upon land in Europe. This water from America may have soaked into European earth, to rise as a spring in some valley where our fighting men are encamped. For water, like air, knows neither frontier nor race, but belongs to all men. . . .

As Emmanuel Parrish and his wife Celista plant their fields, as one Negro figure stoops and scrapes a hole in the earth, and another Negro figure places into this hole the fragile plant of a sweet potato, they obey the impulse to grow things upon their own earth. If they had that vision which can see beyond events of the moment, they would know they are planting for a world outside their comprehension. Fatigue would seem of less consequence, and aching backs of little matter, if they could see where, and by whom, these sweet potatoes would be eaten. For, aided by science, two obscure Negroes upon a truck farm in America are feeding fighters. As they plant their fields, they, too, make history.

# 6: FRUIT

The hucksters' barrows are bright with fruit. They glow against the grey of the city like setting suns. They bring an air of gala to this shabby street, and a riot of color to the food stores of the city.

In the dim light of a winter dawn the markets had thronged with buyers and sellers, and, along the tracks and piers of the railroad terminal, work gangs had broken down the freight cars, loading the cases of oranges upon waiting hand trucks. Into the pale light of morning the auctioneers had chanted their sales, and cases of fruit had changed hands, to be heaped upon the shelves of stores, for the people of the city.

For freight trains have crossed America, with fruit for the people. They have crossed mountain and desert, river and prairie; but the fruit within the refrigerator cars has felt neither cold nor heat, neither frost nor sun. Trains have brought oranges east from the Pacific, north from Florida; they have brought grapefruit north-east from Texas. In the heat of summer they have crossed the sizzling continent with apricots and plums from California, and peaches from the South. Within dull yellow freight cars lie the harvests of orchard and grove.

The orange groves of California are veiled in curtains of smoke from the smudge fires that protect the fruit against risk of frost. Along the tracks of the sorters flow rivers of gold. Around the trees stand filled picking boxes. Woven into this pattern of color are harvesters, hidden upon ladders in the thick foliage of the trees, or bending to empty their oranges into the waiting boxes. Over everything is the sweet scent of the fruit.

The orchards of California have reached their moment of harvest. In the Santa Clara Valley, the apricots lie in gold masses at the foot of the trees,

like pools that mirror a sunset. Over the dusty highway of U.S. 66, trucks bring the pickers to the burdened boughs. But these are not people who have known the trees through budding and blossoming, or in the time of pruning, when they are bare of leaf. These are alien migrants, existing upon the shift of the earth's fruit harvests, and there is little place within their hearts for tenderness towards the orchards.

And yet, as they carry and swing the ladders from tree to tree, delicately balancing the topmost rungs against the boughs, that they may do no damage, these people who tend the great impersonal harvests of California have retained in large part the beauty of all harvesters. From the days of Virgil's Georgics the dance-like pattern of fruit gathering endures; for rhythm enters into all labors of the earth. So, as these impoverished migrants from Oklahoma and peons from Mexico strip the trees of the Santa Clara Valley, their bodies move with the beauty of Italian bodies in ancient Tuscany, or Provençal bodies at the harvesting of the grapes.

In the peach orchards of Georgia and Delaware and the Carolinas, the poor whites and the colored pickers know the shapes of the trees with no more intimacy than do the migrant harvesters of California. To them, as they gather the fruit in the heat of summer, these peaches are potential food and shelter and clothing. The orchards are too vast to arouse affection. But among these harvesters are some who nurse the memory of a solitary peach tree throwing upon the silvery wood of an unpainted cabin, in early spring, the tracery of its shadows: a peach tree that they have planted and tamed and served and, serving it, have loved. Something of this sense of servitude to a tree is communicated to the picking crew, and creeps into the hands that lift and snick the peaches from the boughs.

Servitude it is, over the months of the year. Behind the bushel baskets of fruit that fill the cars of freight trains, or the trucks unloading at the canneries, behind the heaped peaches in the food stores of the cities, stands the epic of disciplined servitude. It is valiant servitude, that can be rendered purposeless by the caprice of the sky. One night of frost at the time of blossoming will defeat months of labor. The fury of a gale can whip the peaches from the trees at the moment of gathering. And a hail storm will cut the fruit to the pits, and destroy a crop within the few minutes it would take to walk from the house to the orchard. But hopeful servitude endures.

They are like hands uplifted to the heavens in supplication, these acres of peach trees that have just been pruned. Over the red earth of Georgia they stretch, and the red earth of the Carolinas. They stand upon a bright green carpet, for the spears of the winter wheat, sown between the rows, have pierced the earth while the trees were yet bare of bud.

Between the rows the tractor pulls the sprayer, and the bare trees, shining silver in the winter sunlight, receive the asperges of the dormant spray. Again and again, through the season of budding and setting, the tractor will chug along the aisles between the trees, spraying against the enemies of the fruit: dormant spray in early spring, against peach-leaf curl and San José scale; blossom-time spray when the trees cover the land with a blanket of sharp pink, and the spray dims the branches and trunks with a mist of unreality, till we look upon the orchard of a dream and wander among spectres on a shadow world; shuck spray against rot and curculio, when the petals have fallen to the ground and the fruit has started to swell. But there are enemies like the peach borer that defies the spray, and the oriental peach moth that resists all efforts to destroy it by poison, for with unerring instinct it spits out the first bite upon the sprayed peach skin; this enemy can be destroyed only by a parasite which lays its eggs within the body of the moth, so that the tiny maggots may feast upon the entrails of their host. The beauty of a ripe peach is the result of ugly battles against the balance of nature.

Looking at a peach orchard in blossom against a blue sky, we do not see, between us and this beauty, the phantom shapes of fruit moth and curculio. Neither do we reflect that this blue sky later holds threat of drought. For the heavens can join forces with the insects in their war with man, and the earth be strewn with untimely fruit, withered for want of rain.

As freight trains carry fruit across America, so, also, trucks rush over the highways in early fall, leaving behind them upon the wind the sweet scent of apples. They come from the orchards of the Hudson River and Upper New York State; they drive north from the valley of the Shenandoah and the slopes of Virginia: and within their bodies are piled boxes of ripe fruit.

On the slopes of a hillside in Virginia stand seven thousand apple trees. The owner who planted this orchard walks among his trees with pride. As a sailor will love his ship, with like intensity will a man love his apple orchard. It is an exacting mistress, leaving him no rest. When she beckons, he obeys, and whatever she may demand, that, willingly, will he give.

But not only love for his trees enslaves the apple grower. He is bound to them by a sense of benign power. With the planting of these trees he has changed the contour of the American land. What was once hill is now orchard, fretted against the sky. He has made a landscape to endure over the years, more lasting than the annually changing garment of barley field or corn. For the field of grain passes with harvest, but the apple orchard is this man's offering to the future. Walking his orchard, he remembers when he had planted these trees, thirty years ago. They were small then, and distant one from the other. But, over the years, earth and rain and sun had worked their concerted power, till the rows seemed to draw closer together, as the trees broadened and grew. His mind retains images of the growth of his trees, that he may look back with delight upon their changing shapes; each tree with its separate identity of design, despite the uniform symmetry aimed at by pruning, each tree strong of form as it springs out from the ground, each tree throwing its blossoming arms to the sky in ever widening stretch year after year. And the spirit of the knight errant enters this man in his battle against codling moth and scab.

He walks the miles between the rows, knowing the shapes of his trees with the peculiar awareness of devotion. And because of this devotion he resents the intrusion of the picking crew. These people rob him of his solitude among the fruit. They rape his trees, these nomad harvesters; and as the branches spring high into the sky, released of their burden of apples, he feels the sting of jealousy. If he could share with them the awe he feels towards his orchard, he would not mind their shouts and laughter among his trees, as they despoil them only for money. But as they scale the ladders and fill the wide-mouthed canvas sacks, strapped in front of them, with crimson apples, their minds are far from the beauty of the fruit.

Blinded by love for his orchard, this apple-grower denies identity to the pickers. To him they are people who neither sow nor plant nor cherish nor tend, but follow the seasons up from Virginia and the valley of the Shenandoah, up through Maryland to Pennsylvania and New York State. They are rootless pickers, who live upon the shifting harvests of the country. Were he to wander among them, or hear them talk at night in the privacy of the pitched tents, he would see that they have but one thing in common; and that is poverty. For one man dreams of a farm in Georgia where the red earth left his land, and gullies stand in the place of cotton. Another thinks of a smoke-covered city in the heat of summer, and a family of half-starved children who would snatch at one of these apples that lie, slightly imperfect, in the grass beneath the tree. That wizened man emptying his sack into a picking box wonders how long it will be before he can gather fruit that is truly his own; his hands touch the apples tenderly, for he knows that the first tree he will plant, when he has bought his land, will be an apple tree with fruit like this. The woman in the torn blue skirt sees, between herself and the orchard, a sick mother in a mining town of West Virginia, where there is little fruit. But all the apple-grower notices is a gang of transient harvesters, plundering his trees.

There is nevertheless a feeling of carnival among these pickers. For man abandons his restraints before the ageless urge of the harvest. A bacchic excitement overcomes him, and, whether he is a Negro cutting broccoli upon a truck farm in Florida, or a nomad harvester among the foliage of an apple tree in the Hudson Valley, or a Catalonian peasant shaking ripe olives from the thousand-year-old tree in a grove in Majorca, something beyond reasoning rises within him, and he needs noisily to rejoice. They swing the ladders, to place them against the trees, these pickers in a Virginia orchard; and, along the thirty miles of avenue between the trees, ladders are swung in identical manner, swaying in the wind, until the movement is like a great fugue.

Were we to see the orchard this morning, when the dew is barely dried from the thick grass beneath the trees, and the shadows have scarcely withdrawn the interlacing of their fingers between the rows, were we to see it at one moment from above, as God must see it, we could understand fully the beauty of this design of pickers upon ladders, pickers clearing the lower branches of the trees, pickers unhooking the flaps of their canvas sacks to let the crimson fruit slide carefully into the waiting boxes. We would behold the filled boxes, stacked three high, dotted over the orchards like cases of jewels. The scent of ripe apples would rise to our nostrils, drawn skywards by the sun. But if we were God in the heavens, we could fling our eyes beyond this orchard and perceive apple harvests flaming over the land. We would see a chain of harvests beyond the Rockies, on the coast of the Pacific, where the pickers in the valley of the Columbia River swing their ladders against their trees upon the grey soil of Oregon with the same dancing movement, and fill the same picking boxes with the crimson fruit: King Davids, Jonathans, Winesaps, and Delicious, each apple true to its own color and shape, they scent the air of Oregon, out beyond the breakers of the Pacific, till the wind merges their perfume with the salt of the ocean.

The apple grower upon the slopes of a Virginia hillside watches the pickers at their work. He searches the filled boxes for evidence of careless picking, noting the bits of twigs that have been left on, the pulled-out stems, the bruises, the broken skins; for his apples must arrive at their destination perfect in condition. In the packing house, sorters sit at high tables, and the apples are graded by size; they travel along moving belts, and machinery brushes and polishes them, till, as they are packed in the new-made boxes, they have the impersonal beauty of perfection. It is as though they had renounced their origin—for who could imagine one of these shining things moving in the wind, being burnt by the sun or splattered roughly by a shower of rain? They have forsaken their earthy heritage to enter civilization.

But their grower is not deceived by this new splendor. He can see those apples when they lay within clusters upon bare branches, before the buds had even separated, and the thin mist of the pre-pink spray covered the trees. He sees them when the pink has started to show at the tip of the blossom, and the tractor draws the sprayer between the rows for the pink spray. He hears bees in the blanket of blossom, and praises the sky for its gentle warmth. For the destiny of an entire apple orchard for one year depends upon the weather during the week of blossoming. Man may build his cities and win his wars, but his power must bow before the honey bee in an orchard. He may prune and spray and use the latest findings of science, but let the week of blossoming be such that it does not tempt the bee from the hive, and his trees will flower with a sterile beauty. Let the wind blow cold, or the sun hide behind grey clouds till the rains fall, the apple-grower will walk his orchards in distress, pausing beneath the blossoms to listen to the stillness above him; for there is no hum of bees in token that his trees will bend low this season with fruit. Within their hives the bees stay warm, secure from battering winds; they neither know nor care that their ease will cost a man his year's livelihood. The destiny of his orchard lies in the hum of bees among the blossoms.

As the apple-grower looks at the perfect apples within the boxes, his mind follows the development of those apples over the time of growth. Circles of shade appear under the trees, as the foliage thickens, and little green bullets of fruit grow among the leaves. And then, when the tiny apples begin to swell, they crowd each other upon the boughs, and ladders are swung against the trees. The orchard dances with the movements of the thinners, stripping unneeded fruit. The little green bullets drop to the ground with the sound of hail. The grower sees a kaleidoscope of trees and men, ghost-like from cover sprays, and hears the hiss of the high-pressure jet beating upon the leaves. He searches the molasses and sassafras traps for the dusky little codling moth that can wrench his harvest from him. He watches the sky for rain, remembering droughts that had withered his apples; for he can feel within the cup of his hand the shrinking fruit from which the mother tree had withdrawn the juice, to send the moisture back into the leaves, that the tree might live.

He looks down upon the packed, polished apples with pride upon his face; for he knows the enemies that he has fought, and the hostages he had given to the skies.

The apple harvest is ended. A strange quietness falls upon the orchard, broken only by the occasional thud of ungathered fruit, falling to the ground. The birds have ventured back to rest upon the trees. The pitched tents of the migrant harvesters have disappeared. Down from the trees drift the first dead leaves of the autumn. Over the states of America, apples from these trees are travelling far in their wooden boxes, in freight train and truck; but the trees themselves will stand leafless soon, looking ravaged and desolate, as though of little worth.

The man who planted these trees walks along the shadeless aisles with contentment on his face; for he knows, as he looks upon the bare boughs of these trees, that already, within this dead-looking wood, buds are forming for next year's blossoming, and that the roots beneath the soil are drawing up into the substance of the trees power that will turn into globes of crimson fruit. This man who has given his freedom to an orchard of apple trees knows that the cycle of the year is never ended, and that at no one moment does growth cease. He has found a worthy, if exacting, mistress.

## 7: MILK AND EGGS

Sometimes, as we drive along the highroad, we overtake, upon the upgrade of a hill, a great white truck. It is a glass-lined tank, filled with milk for the city. If we are early enough in the morning, we pass farm trucks loaded with milk cans for the country plants, where the milk is weighed and sampled and cooled, to be put into forty-quart cans, or into these tank trucks, for transportation to the big city pasteurizing and bottling plants. And if our road lies beside a railroad track, we may probably see a lengthy milk train carrying its refrigerated tank cars to the cities. For milk, that most perishable of foods, must be handled with promptness and care. Here is no slow freight of grain that can wander at leisure across the country. The people of the cities must have their morning milk.

Over the ice-bound roads of winter go the glass-lined tanks, over the fiercely hot high roads of summer; but within the tanks the milk stays cool. We have conquered time and space.

But if we are able to transport fresh milk across a continent, there are some things still beyond our control. Upon the dairy farm the rhythm of life is unchanged by our efficiency. Here cows stand and stare as they have always done, and even the mechanical milker cannot hurry the creatures in their stanchions as they chew their cud.

A river of black and white flows over the meadows. It is a slow-running river, meandering among the field flowers of late spring. A herd of Holsteins is being driven to the milking, and will not be hurried. The cows pause, and lower their heads to the lush grass for last mouthfuls of this sweet food; memory has stamped upon their dull minds the long months of winter, when existence was a repeat pattern of day and night, in the dim light of the barn, and their sole drama lay in the letting down of their milk, morning and evening. Were they conscious of the elements of time and season, they could have solaced themselves, over the lightless months, with visions of sunny meadows. But whatever was happening within their minds was hidden from the cowman, who saw only long rows of Holsteins in the stalls, and did not question any thoughts behind those gentle eyes, so long as udders were heavy with milk.

They loiter now, as they are driven to the barnyard, in what must be delight. This is the turn of their year. Now is the beginning of summer, with long days in the sun-soaked pastures. Today they have left the closeness of the barns, and the taste of the grass is sweet.

It is no wonder the cowman guides his river of black and white with difficulty towards the barnyard. But at last they are out of the meadow, wandering along the dirt lane that leads to the farm. It should be easier now to control this flow of cattle; but still they linger, swinging their tails round to the massive bodies, as they flick at the flies. By the side of the lane grow patches of clover, waylaying the cows; they crave this pasture with the longing of sun-starved creatures. They walk with awkwardness, their udders heavy with milk. Look closely and you will see the milk veins standing out on the bags, and the teats hanging vertical and pink. There is no force in this world to curb the need of a cow to be milked. Morning and evening, summer and winter, the farmer is in bondage to his milch cow. This is a necessity he cannot gainsay. He may look at his fields and decide that tomorrow, or even the day after, he will harvest his wheat. He can delay in the sowing of his seed. But though the world should collapse around him, the udders of his cow will fill with milk. He is slave to his cow, for she cannot wait.

Up in the mountains a man courts his girl. He would stay with her until the shadows lengthen down the hillsides, and evening falls. But across this mountain, in the next cove, his cow is waiting. Before him he sees the fullness of her udders and the look in her bovine eyes. Stronger than the urge of his courting is the call of his cow. He must leave this softness among the hills. His cow must be milked.

But the Holsteins have not yet reached the barnyard. The cowman is gently patient; he has herded these creatures over many years. The dirt lane crosses a stream, and the sun is hot. The stream runs full, from the rains of spring, and the cows are warm from the sun. Into this stream they wade, straying from the path to the farm, and the low-hanging branches of the willow trees brush the backs of the cattle. Slowly they wade in the stream, unregarding of time, and the cool waters lap their heated udders and the little curls upon their bellies. They drink, and great circles overlap on the surface of the water, with intersection and curve, as cow nears cow. They flick the wet ends of their tails over their backs, and the drops of water sparkle in the sun. This is the gentlest of scenes, and the more comforting because it is so completely benign. There is not even the spectre of the slaughter house behind it. Now, with the world at war, there is needed solace in this beauty, for we look upon an assurance of tranquillity, and know that gentleness remains upon the earth. In an age forced into destruction, the quality of gentleness is apt to be confused with weakness, for violence becomes the unit of measure. It is for this solace that we linger now among the cows, watching their softness with delight. For we know we watch something that will outlive the fury of wars. Cows will be brought to the milking when battles have been forgotten and dynasties vanished. They will linger in cool waters across the face of this earth, and patiently they will switch their tails over the straight line of their backs to flick off American flies, English flies, German flies, and flies of all nations.

The cows gather in the barnyard. But it is not yet the moment for milking. These Holsteins must wait. For it is spring, and the pastures today are tender and young. The cows must rest for a few hours, that they can belch out the grass taste. The milk, else, will taste of this grass. Here in the barnyard they stand, chewing the cud of the lush young meadows. As the year rolls on, and the grass grows more tough, they can be left in the fields until milking time; the river of black and white will flow over the land with a sun low in the sky, casting the shadow of each animal before her, and binding cow to cow, till they flow as one body.

The afternoon draws on. Now it is milking time. The farmer opens the doors of the barn, and the cows file to their places, each to her own stanchion. But there is one cow who will not follow. She has bawled through the hours of today, crying for the calf that was taken from her. Wild-eyed and lone, she calls for her child; but it does not come. The farmer tries to coax her, but her desolation is stronger than the beguilings of the herdsman. She roams the barnyard, her head high to the sky, bawling for her lost calf.

In this modernized dairy farm, the cows are milked mechanically. But first the udders must be washed, that the milk may be clean. If this scene had taken place in the winter, when the cows are kept in the shelter of the barn, the animals would have been fed. Down from the two silos would have come the cut corn, smelling sweetly of fermentation. For the milk we drink in the lightless winter months is the harvest of fields of corn. Over the dairy-farm country the corn thickens and grows tall. But this is not corn to be left standing upon the earth until the green has withered. This corn is harvested while goodness is still within the stalks. Across from the little stream where the cows dallied, the corn is being cut. The harvesters haul the juicy plants to the silos, at the end of the barn. And there, in the hot days of late August, to the accompaniment of belts and conveyors, and the chug of the tractor, the corn is cut and elevated to the silos. Into these solemn forms that stand sentinel against the barns of America, go the sun and moisture of the cornfields, and the goodness of this earth.

This is the harvest that is fed to the Holsteins who stand, throughout the dim months of winter, in the shelter of the barn. Rows of black and white cows turn their heads slowly towards the farmer as he rolls in the truck of silage, and let down their milk to the soft sound of eating, until the barn is filled with a subdued ripple, as of water flowing gently over rocks.

But it is summer now, and the cows have been feeding all the day, upon sunny pastures. They are ready for the milking, their udders heavy, the milk veins swollen upon the bags.

Of all the "iron men" that work upon the farm, needing neither food nor rest, the mechanical milker seems most to lack beauty. It enfolds the teats, pulsating with a robot-like movement. We watch this with a strange sense of distaste, even while we acknowledge the inevitable. Seeing beauty in the tractor upon the fields, and in the corn-picker that devours the ears with such mechanized greed, we question this feeling of distaste. Why do we lag just here, when we have accepted other forms of the machine? We have no answer, but, as the men strip the cows, following in the wake of this unhappy-looking mechanical milker, we see something deeply satisfying in the design of human against cow. Is it aesthetic pleasure we demand? Or is there something in the communion of man with his animals which gives us contentment? It is easy to imagine that this cow he milks is pleased by the touch of the man's head against her flank, and that she feels some sweetness in the clasp of his fingers around her teats. And we like to believe that the milker is warmed by this closeness. At this moment they are kin.

**Over the countryside,** in late afternoon, farmers milk their cows. If we could see all of this land, at one moment, we would look along lines of cows in their stanchions; black and white Holsteins like the painted wooden toys of a child's farmyard, red and white Ayrshires, warm fawn Guernseys, sooty little Jerseys with their short wide heads and protruding brows, and "just cows," as the little farmer will call his medley of animals. A mechanical milker would be fixed to several cows at a time, in the great barns of the dairy country; and on small farms we would hear the thin spurt of milk hitting tin pails, and a low grumble from a milker when his cow shifts her balance, or leans the weight of her body too heavily upon him. We would see farmers with the peaks of their caps turned backwards, that they may rest their foreheads with ease against the flanks of the animals, and wives and hired girls in cotton dresses and sunbonnets, sitting on three-legged stools for the milking. The farmers' sons would be there, and the older children, just home from school, while a swarm of cats would appear, waiting for milk; they would brush against the legs of the milkers, and sit on the backs of the cows.

It is the most timeless of scenes, filled for all men with recollections. Perhaps this is why we cling to the image of hand milking. We went to the barn as little children, to visit the cows. We looked at books pictured with sunbonneted milkmaids, and sang nursery rhymes about them. We gazed upon cows knee-deep in buttercups, in the month of May. We saw them coming home down the lane, and whatever country it may be, they linger in the same manner, whisking their tails with identical curve.

But this milking that seems so idyllic on a summer afternoon holds less enchantment in the bleakness of winter. For, as dawn nears, the snow and the cold turn the short distance between farmhouse and cowbarn into a feat of endurance for the shivering farmer.

It is no wonder that the barns of New England are part of the dwelling-houses, as protection against the rigors of their winters. Today, in January, this farmer would give much for a mechanical milker. His hands are numbed with the cold. He can scarcely feel his fingers as they clasp the warm teats. There is comfort in the heat of the cow's flank, and he rests his head against her with pleasure. Into the tin pail spurts the milk, foaming and warm. But before his eyes the farmer sees the long drive to the milk plant, over icy roads on the ridge of the hills, to the outskirts of town. The snow has piled high in the night; it is still falling. But the people in the cities must have their milk. They must have their butter, and their cheese. Over these icy roads the trucks must bring this milk from the little farms.

In the early hours of a winter morning, trucks converge toward the plant, and the farmers help each other to unload the cans. They meet, and their talk is of ice and snow, and the sickness of animals. And talking so, they are full of friendliness towards each other, until the biting air of the winter morning seems warmed by their spirit. Here is no rivalry. Here is the will to help. Does a cow have an inflamed udder, a farmer will suggest a certain bag balm. Has one man an ailing calf, they will draw closer into a circle, comparing remedies. For each man of this group fights the same foes, and carries within his life identical hostages.

It is good to feel this human warmth here, at the milk plant. These men who bring their cans over the wintry roads have milked their own cows and helped at the birth of their calves. They have planted and harvested their own corn to feed the cows, and have sat in the barn through cold nights of winter, watching a sick animal. These are not men paid to labor for others, who work with their arms but not their spirits. These have dedicated themselves to their cows.

The air inside the plant smells sweet, like the breath of a cow. It is a world of revolving machines, and pipes and conveyors, of steam and noise, bottles and flowing water. And as we look at this pasteurizing plant, and watch the empty bottles being filled, we feel the beauty in all these movements, and know that this is the dance of milk.

There is yet another harvest of the land. If it seems to lack magic, this is only because we are adults; for a sense of wonder before the common happenings of life fades with childhood. This harvest is the miracle of eggs. In the farmyards of the world, hens scratch for food, and farmers' wives scatter corn at feeding time, with the gesture of a sower upon a field. For the chickens seem by ritual to be the woman's domain, and it is the wife who keeps the egg money. Chicken-houses stand over the land, and wire fences enclose White Leghorns and Rhode Island Reds, Wyandottes and Orpingtons. For America must have her breakfast eggs. She must have her chicken dinners.

The hatchery is on the edge of town. You can hear the cheeps of the day-old chicks as you enter. They are piled in boxes, ready for shipping, and through the ordinary parcel post go these fluffy atoms, to states over America, fortified within their tiny bodies by seventy-two hours' nourishment from the yolks of their eggs.

Farmers' trucks are crowded at the doors of the hatchery, bringing hens' eggs from miles around. Flocks of hens over the country are laying eggs for this one hatchery, and a chicken growing to maturity in Missouri may have been hatched from an egg that was fertilized by a rooster in Kansas. The eggs are placed into trays; they pass now into the realm of science. At the end of five days strong blue lamps will be placed beneath them, to determine whether they are fertile, and the trays will be slid into incubators, where the eggs will hatch out in twenty-one days.

It is a strange place, this "forced draught" incubator; it is oddly different from the feathered body of the mother hen. A warm wind blows through the impersonal corridor, and in rows of trays, on either side, fifty two thousand eggs are working their miracle of birth. Already, if we listen, we can hear the feeble cheeps of a few baby chicks that have broken from their shells. Let us draw out one of these trays, and watch this miracle. They are pipping in the tray, and we can see a hole in a shell, disclosing the inside fibre. Several chicks have hatched; they stumble among the eggs, some of them downy and soft, some clammy still, from the shell, with a strange silky look to their wet bodies; these are shabby little creatures, as they stagger around. To the right, at the end of this tray, a baby chick wriggles to free itself from the shell. It is weak, and the tiny head hangs limp, to one side. But this creature must not be hurried in its struggle for birth; the membrane that binds the chick to the egg must not be broken, lest the insides of the little bird be tugged from it, and it die. Now it has emerged, for the membrane is separate, and dry. Over the surface of this tray of eggs, tiny beaks are pipping at the shells: first a little hole, then a crack running round the shell, then the entire top is opened, disclosing the chick, curled snugly within. Tomorrow this tray will be alive with hatched chicks, and the quietness of the incubator will be broken by cheeps from thousands of new creatures. Little beaks will poke from the spaces between the trays, and tiny eyes will gleam, and when the trays are drawn out from the shelves the birds will overflow on to the floor, so bountiful is this nursery.

In a far corner of the yard, just before you reach the path leading to the bee-hives, the little brown hen has hatched her chicks. She watches them with a fierce eye, as they wander from her reach; for she is imprisoned within the wooden bars of her chicken-house. She follows the movement of this downy flock, anxiety in her look. The children are back from school, and rush to the yard to see the newly hatched birds. They lift them into their hands, and play hide-and-go-seek with them in this grass which is taller than the chicks themselves. The hen cackles in agitation, as the children heap her babies in their frocks, and the farmer's wife comes to the yard to see what is happening. She stands there, watching this gathering of young things, and a softness creeps over her face. At this moment she is thinking mostly about the little brown hen, and while she reproaches her own children for their roughness, she murmurs: "And I'm so glad she's had success this time. She tried so hard, and she generally has such bad luck." This woman identifies herself with the hen, seeing little difference between bird and human. For she knows the look in the hen's eyes, and shares her troubles. We watch the freemasonry of motherhood.

The magic of eggs is still understood.

## 8: FARMS

THE TRAINS THAT CROSS THE COUNTRYside of America with food for the cities pass fields of corn and wheat. They pass cattle grazing upon pastures, and hogs wallowing in the mud of barnyards. They toss their shadows upon fruit trees, as the railroad tracks bisect the checkerboards of orchards, and whisk past long rows of vegetables stretching to the horizon. But they also pass the farms that shelter the cattle and hogs and the harvests of corn and wheat. The shriek of the locomotive whistle muffles the sound of the farmer's wife calling her man to dinner, and the rattle of the cars is heard by hired men who sweep the food troughs of cowbarns.

The lengthy freight train that moves across the country with its burden of grain and hogs passes bursting corn-cribs and filled granaries. The engineer driving the locomotive over the prairie sees valiant little farms, many miles distant from each other, and the lonely workers wave to him as the train crosses their earth. He sees rail fences, like protecting arms, around clumps of buildings, and silos and sheds, farmhouses and barns, grouped in the shade of groves of cottonwood trees. He passes little farms owned by struggling men, and great newly painted "improvements" that are the property of prosperous farmers.

But the freight train goes its way to city and terminal, fading into the distance behind a cloud of smoke from its stack. It does not pause in its passage, to learn more closely what is happening within these farms. It cannot know the pattern of life that is lived among these silos and barns, or enter into the spirit of the farmer who works this land. These clumped buildings house the force that shapes the earth; they are the resting place for the weary, after days at plow. For this farmland whose harvests are carried across America in the cars of freight trains, exacts endless labor. Day after day, throughout the months of the year, the farmer must work with his animals and his earth.

But on the seventh day, in ordained manner, the farmer rests. His cows milked, his cattle watered and fed, he dresses in his best clothes, to walk his fields.

There is a godlike dignity to him as he walks his fields. Over the hills of Maryland and the plains of Illinois, on the undulating land of Indiana men look upon their husbandry and see that it is good. Across the Atlantic, English farmers walk their fields on Sunday, and in the war-stripped countries of Europe phantoms gaze upon scorched wheatlands and trampled vines. For it is an instinct older than recorded history, that man should pause to look upon his handiwork. And as he walks his fields, he is filled with undefined humility, feeling himself of no greater importance in the scheme of life than the bee he raises from the clover or the intruding mustard he pulls from his alfalfa. Feeling this, he knows that in sowing his seed he has liberated forces beyond his control. Often stubborn, sometimes hard, never is the farmer arrogant; for he knows that he is servant and not master, and that one storm can destroy the labor of months, while in the slow despair of a drought he can be defeated.

Below him stretches the great valley. It is quiet today, for tractor and team rest within the barns. But the mind of the farmer leaps this quietness and places a plow upon the curve of the hill, to the right of the clump of locusts. It covers the pasture on the top of the first rise with the black and white pattern of a herd of Holsteins, and draws the harrow across the waiting earth to the left, past the little stream that wanders among the willows. It turns this field to corn and that distant one to sweet clover, and shifts the balance of color upon the great quilt of land below him.

Looking down upon this powerful valley, it is as though the farmer were gazing upon the face of someone he loved. He knows that face when the sun shines on it and each feature—pasture and plowland, silo, farmhouse and barn—glows in the light. He knows it when the shadow of a cloud crosses its brow and darkens the field of barley. He knows it when heat lies upon it in a mist, and the mountains are unsubstantial as gauze, and even the gold of the ripe wheat is dim. He thinks of it on a morning in early fall, after the first night of frost, when each color stands sharp and clear. Sunrise and sunset, rain and wind, he can see it in all moods and seasons. And as the lover watches for the least sign of distress upon the face of the loved one, so the farmer seeks for a sign of discomfort upon the face of his earth; his keen eyes search the fields for ditches that must be cleared, and pass slowly across the roofs of his farm buildings, lest repairs should be needed.

Across the valley, away to the distant mountains that shelter this earth, clumps of farm buildings lie spotted upon the land, tucked among clusters of trees. Seen from this hillside, they look in the distance like toy dwellings, and it is not easy to remember that each clump pulsates with the life of a farm. Within each minute building is a family that went to church this Sabbath morning and praised its God. The hot smell of food came from a kitchen in each farmhouse, as a woman cooked the Sunday dinner; and now, in the lazy warmth of early evening, these farmers' wives rest their legs upon horsehair sofas, in the stillness of parlors, and hired girls deck themselves in their smartest clothes, dreaming of movie stars. Each group of buildings holds within its walls the range of all human drama and all possible emotion. Across the valley lies one old man who will never again walk his fields; he gazes from the window of his bedroom upon the land he has held and worked in trust for his son, and as he hears the sudden bleating of frightened sheep, he strains his eyes towards the right, past the cowbarn, where his little grandson must be chasing the lambs. The old man lies back upon his pillow, knowing he has served posterity and that life has been good. Within another farmhouse, nearer to this hillside, a woman rocks her baby; she rocks the fourth John Mowery to have been born among these fields. As she puts her baby to sleep, she watches a tiny black speck moving across the rust-red serpent of the contour plowing that is planted to corn. The black speck stoops, as though to examine something upon the earth. It pauses and, lifting its head, looks far away, beyond the barley field and the winter wheat, to where the mountains touch the sky. The black speck is John Mowery, the father of her baby.

Over the valley this Sunday move many black specks. They are so small upon the enormous countryside that they seem of little significance. But they are the thinking, sentient powers that shape this earth. Without them forests would cover the land as in the days of the Indians, and a tangle of undergrowth would choke the pastures and strangle the structure of silo and barn. They are the forces that design this valley and make of it a scene of great goodness. "Tomorrow we will plant the corn in the twenty-acre field," thinks one black speck, as he stands upon a small rise overlooking his farm; and with this thought he stamps upon the picture of his land lines of bright green, then twenty acres of buff tassels, and, in the flaming days of the fall, an encampment of corn shocks bivouacked across the valley. He fills the white silo against his barn, and patterns his pastures next spring with fattened cattle. "Come fall, I must put that cornfield to clover," decides another small speck upon a further hill; and, with this decision, unborn bees are given the promise of a harvest of honey.

The quiet that lies upon the valley this Sabbath is not the quiet of a dead stillness. On this seventh day man may rest from his labors, but the grain fields ripple in the wind, like yellowish green water. As we look down on these fields, it seems strange that the seas of barley should stay within their bounds, for this movement of the wind among the heads might mount in a great wave at the fences, to flood neighboring fields. But the soft barley ripples within its confines, endlessly swaying. And then, as we turn towards the left, where the fields creep over the hill, this green sea of wind-blown barley behaves counter to the ways of water, and flows upwards. It flows uphill, but never encroaches upon the earth of the plowed fields, so static against the swaying of the grain.

The farmer gazes upon this goodness. If his walk were to take him higher, to the summit of the nearby mountain, where he could look down upon the buzzard in the sky below him, he would see more fully the beauty of his valley. He might think of the words in the Book of Malachi:—"for ye shall be a delightsome land, saith the Lord of hosts." A sense of awe would fill him, as his eyes fell upon those fields that were his, until ownership became a loyalty and a trust, not only to fields and sons, but to all mankind. For the rich patterning of this earth beneath him—grain field and pasture, woodland and orchard,—lies in his hands and in the hands of fellow-farmers. From this height among the blossoming bushes of mountain laurel, he would see where science had started man on his conquest of wind and rain. Had the father of this farmer been brought back from the grave to look upon the panorama beneath him, he would have wondered at the strange curves winding over the land. No fields were plowed thus in his day. But the farmer who walks his land this evening in May knows that the sensuous curves of contour cultivation are more than exciting beauty over the landscape; they guard the topsoil upon his fields, that wind and rain may not wrench his goodness from him. Even here, on the lesser heights of his own hillsides, he can look with content at the great green serpents of clover that wriggle across his land, and the rust-brown serpents curving upon the slopes of his hillsides, planted to corn. A new pattern has been given to the American earth.

But looking upon his fields, on this day of rest, he knows that for the farmer there is never rest, and that his harvests will come only from unending labor. This picture below him is not made by walking the hillsides on a sunny evening in May.

Thinking this, the farmer sees himself before dawn on a morning in winter. It seems as though the rain never would stop. It beats against the bedroom windows, and the wet cold air strikes him as he gets up from his bed. Outside in the yard the mud is ankle-deep. He gropes his way to the barn, and the sheep, hearing his steps, begin to bleat. They are there as he opens the barn, anxious for their feed of oats and cracked wheat, the lambs by their sides.

He pushes back the doors of the stables, leaving behind him the comforting sweet scent of hay; for the air within the stables is thick with animal smell, and the smell of ammonia. There are two sick calves here, and the place must be kept closed and warm. . . . It is good to throw open the doors and let in the air for a moment, though it be raining and cold. . . . As he moves through the building, in the blurred light of dawn, twenty-four Hereford steers turn their white-faced heads towards him, with identical movement of back and shoulder and neck. At the other side of the stables his Guernseys look his way, soft-eyed and gentle, and the new Angus heifers swing themselves round to gaze at him, their black bodies scarcely visible in the dim light. His animals greet him, with quiet, heavy salutation, waiting for food.

He must clean the troughs before he can feed the steers. He climbs into the pen, with the animals, and the warmth of their breath is welcome as he pushes the great bodies out of his way, so that he can sweep the troughs. He would linger among the animals, warming his numbed hands on their flanks; but the hired man is sick, and there is much to be done. He fills a bushel basket with ground corn and spills it along the trough; in their hurry to eat, the impatient creatures plunge forward, and receive a sprinkling upon their heads. They rush at the corn, three squeezing into a space intended for two, and the sound of their feeding is like heavy rain upon a roof, or water running over rocks. . . . But the farmer has little time to pause and enjoy his animals today. His Guernseys stand waiting, their udders full. . . . Above the sound of the feeding of animals comes the squirt of milk into a tin pail. . . . Upon the roof the rain beats down, and the world beyond the stables is hidden in mist.

It is strange to live back to those winter mornings now, while the farmer looks at his fields this evening in May. The sun is low in the sky, gilding the land. Over the hillside he strolls, making for home. He pauses at the barn. Against the sunset it stands, on the top of a little rise. The immensity of its size dwarfs farmhouse and cattle, till they look like toys in the nursery of a child.

Up the grass bank walks the farmer, and it is as though he crossed a drawbridge; for this bank-barn is a fortress defending his harvests against the enemies of wind and storm, heat and cold. Rain may drench the countryside; but within the barn the threshed grain, the husked corn, the baled hay, is secure.

Inside the barn it is no longer like a fortress. There is a consecrated feeling here, as though we were entering a church. The narrow windows remind us of a church, and the over-sweet scent of silage might be incense, heavy in the air after the singing of the mass. Darkness surrounds us, with slits of pink light from the sunset filtering through the windows and cracks in the walls. This light falls upon baled hay and alfalfa, and the forms of binder and tractor. Let the sun be higher in the sky and it would cast bands of deep gold across the heaped corn in the crib, and enrich the dull red of the piled cobs; it would dazzle in a confusion of small suns upon the metal of the machinery and slip its light beneath the doorway of the barn, to burnish the combs of stray chickens that scratch for grain upon the haymow floor.

But in the darkness, here, it is sounds we notice. An animal whines. We listen, and then, as we walk further into the barn the whining grows louder. This is no animal in distress; this is the wind forcing its way through crack and crevice. Outside in the fields, the breeze ripples freely in the barley, but here, within the confines of the barn, it moans like a small hurricane. Above our heads, in the rafters, pigeons coo. Hearing footsteps, they fly past us, through the open doors, with a whirring of wings. Among the sacks of ground meal, on the floor of the barn, cats scurry after mice. In a pause of the wind we hear muted animal sounds from the floor below. A work horse stamps her foot; a sheep coughs; a lamb bleats. But even if we did not hear these sounds, we would know that cattle and horses and sheep rested below; for into the over-sweet smell of the silage and the dusty scent from the haymow rises the warm smell of stabled animals and trampled straw.

Let our eyes grow accustomed to the darkness and we can pick out the forms of the great beams that support this building. Mellowed by age, they, too, have the structure of a church. And over to the right, past the canvasses of the binder that hang from the rafters, there is a wing to this barn, as large itself as a village chapel.

Unwittingly we tread softly, for there is hay beneath our feet. But something makes us lower our voices in speaking, as though we stood in a holy place. Within this barn of the Pennsylvania Dutch country, this immense crimson bank-barn with overhanging eaves, here they may actually have held their religious services, these God-fearing Amish farmers. And there seems rightness in this, for the building is custodian of all they hold most dear. It is the token of harvest, and the guardian of their grain. It is a symbol of love for the earth, and of that love which is at once humble tenderness and a sense of mastery.

Over the American land stand barns: crimson barns tucked into the folds of the Maryland countryside, white silos sentinel against them; immense barns, sprawling like prehistoric creatures upon the flat earth of the Middle West; clean-shaped white barns of New England, and the dairy farms of Wisconsin; barns in the rich land of the Genesee Valley, with the name of the owner and the date of the farm's establishment painted proudly upon the front of each building. Over the face of the earth stand barns: thatched in England, with the patina of centuries upon them; small and fragile in the Austrian Tyrol, where they are hipped to reject their burden of the mountain snows.

Always, in looking upon barns, we are aware of the same feeling: that here, in a world of destruction and unrest, in an age of ugliness and change, we can find comfort. For we look upon the eternals of husbandry. Within this great dark barn we are beyond frontiers, politics, and parties. We are surrounded by things that endure: seedtime and harvest, and the tillage of the soil. And we see, too, those things that must be restored to the world, so that once more the peasant in Europe may fill his barn with his own threshed grain and hear in his ears the gentle sounds of contented cattle. For a world that is happy and at peace is a world of well-filled barns.

The farmer who loiters in his barn this Sunday is not thinking all these things. He does not fling his mind across the seas to fellow-farmers who speak languages beyond his understanding. If he were suddenly, as on a magic carpet, to be translated to an English barn in the Vale of Aylesbury, he would feel bewildered by the thatch upon the roof, or the strange piles of golden swedes that catch the sun even as his own golden corn catches the sun. But let him wander for a little while in this English farmyard, in the shelter of the haystacks, in the protection of the sheepfolds, and unconsciously he will begin to understand the universality of the earth. The same lambs butt their dams, the cattle make the same sound as they eat, and though his English brother may never have worn overalls, yet soon they will discover, these two tillers of the soil, that they are servants to the same master, who have given identical hostages to earth and sky. Let that magic carpet whisk him still further, to a village in the Graisivaudan Valley, in France, where the great white oxen used to pull the plow across the mountain sides, between the vines. Let him talk for a moment (but he cannot talk with him in words, for the American farmer has never learnt this Dauphiné patois), let him speak with his fellow-farmer in that language of craftsmanship that is beyond the need of words, and we shall find that they understand each other—this farmer from the prairies of Illinois, this farmer in the French Alps—as they look at the scrawny little cows in the barn and finger the fodder in the haymow. Let him wander yet further, into Italy or Germany, Poland or Russia, and he will find men who would till the soil even as he does, and gather into their barns the harvest of their earth. For the barn is the merging place for mankind, beyond nation and race. It will outlive dictator and empire, for it is the symbol of earth power.

Into the darkness of his barn moves the farmer, past the long sacks of wool from the new-shorn sheep. At the far end of the building stand his implements. He pauses among them, even as the farmer anywhere on this earth must pause; for man sees magic in the instruments of his craft. In the darkness of the cave in the Dordogne Valley, the Cro-Magnon hunter must have fingered his arrowheads with like tenderness. They stand here in their readiness, each with its purpose, each shaped by that purpose and with the beauty of form of all things constructed for work: planter and binder, rake and cultivator, collar packer, harrow, yokes and trees. As the farmer wanders among his implements, he sees the earth's year telescoped before him. The hum of the binder sounds in his ears at the harvesting of his wheat, and, raising his eyes to the manure-spreader, the cold damp months of late fall creep upon him. As he looks towards the rake, the barn is filled with the scent of clover, and before him stretch plowed fields, awaiting the planter. Within the walls of his barn the seasons unfold.

But it is below, among the animals, that most magic is felt. For this is the world of fertility, and the farmer rejoices. This is the cradle of his young, and he walks down the wooden stairway of the barn with softness in his step, as a mother treads to the room where her babies sleep. Punky the cat has had kittens. The farmer kneels in the straw and searches beneath the hay rack for the newborn creatures. While he does so, he looks up at the Angus calf that was born yesterday. It stands on wobbly legs, frantically butting its mother's teats. Beyond, in the further stalls, Sappho and Daisy are just coming to calf. Thinking of these two cows, the farmer smiles, knowing suddenly why he has this good feeling about his animals; whether you feed them or breed them or milk them, you are giving them pleasure. Out in the pigsty the sow has farrowed; she knew when her time had come, and lay back upon the straw to let the thirteen young pigs shoot out from her, steaming upon the cool air as they shook themselves free from the membrane. They heap upon her now, nozzling the teats.

The farmer wanders from his cattle to the sheep with their lambs. As he leaves the barn he passes the stallion; the great beast quivers throughout his skin, shaking his mane as if he were running upon open fields. Out into the sunset steps the farmer, and it is as though the whole world were created fair: for it is a world whose chief concern is with the fruitfulness of the earth, whether it be barley in the fields, Angus cow or little tabby cat.

# EPILOGUE:
# THE EARTH REMAINETH

THE EARTH WHICH YIELDS US FOOD FOR our sustenance holds powers beyond our control, for it is in league with the elements. And the elements are where we brush against the infinite. Here for one moment we can see where time ends and eternity begins, and wars become a momentary struggle and fortunes a mockery. Up in the heavens a tiny area of disturbance may deepen and grow in force until it can mold our harvests and bring scarcity or plenty to the earth. It can mount in fury until storms rage, and turn our fields into lakes. It can withhold the rain from the earth, and the corn will shrivel within the husks.

For the elements are the one frontier which remains unconquered, and our food is made of wind and rain, sun and snow.

In the mountains, away from the cities, the snow falls thick. Throughout the months of winter it cloaks the rocks upon the summits, but beneath this covering, in the dark below the rocks themselves, waters are working for our use. Into the hard substance sinks the moisture of the snow, into fissures between the rocks, along uncharted subterranean pathways; and springs will gush from the hillsides, or rise in the valleys below, to water the cattle.

The snow that falls over the land means wheat and corn, beef and milk; for it is food for the earth out of which all these grow. The white blanket protects the shoots of the winter wheat from ice and cold and the bite of the wind; when the spring sun shall melt this snow, the young shoots will stand erect and green, powerful for growth, fed by the soft snow waters. Across the land which awaits the planting of the corn, the snow will melt and nourish the earth.

The heat from the sun strengthens. Up on the mountain-tops the snow begins to melt. It soaks the small pockets of earth between the rocks, and pours in tiny rivulets to the little mountain streams, swelling them as they rush to merge into the small rivers at the foot of the hillsides. The small rivers tumble their waters into larger rivers, which irrigate the meadows of the upland country where the cattle graze; and the grass that feeds the beef cattle and the cows thickens and grows sweet.

We are ruled by the air, but because it is invisible we forget this mastery until it becomes evident as wind. And this wind shapes our harvests. It blows soft from the South, thawing the frozen prairies of the Great Plains, and the grass grows for early pasture. It warms the winter-chilled earth, and seeds germinate. It brings cool rain to the wilting crops, and, in the hot days of July, it fertilizes the corn, shaking the pollen from the tassels down upon the silks.

But it is not only benevolent. Wind, that can warm the chilled earth, in its turn can snarl and destroy. It can scorch the corn crop until there is no juice in the ear, and rip the life-giving mantle of soil off the face of the land. The same movement of air can pull the "black duster" out of the west, blowing into the noses and eyes of the Nebraska farmer and his wife, blowing into their mouths, their house, their food, and even their bed, the destroyed fertility of the cornfields. Hot dust, biting dust for days on end, and the stifling lash of the wind cuts their faces and their limbs. It moans in the night, driving the dirt through chinks in the windows. The farmer and his wife in North Dakota spit dust, while over their fields the hot wind shrivels the corn, and they watch the destruction of their labor.

The farmer searches the sky. Day after day his eyes have seen the same dome above him, clear of cloud, burnished with heat. Frustration is upon his face. For where man battles against the elements he is defeated. He turns now to the fields around him. He sees parched corn, stretching to the horizon. Behind the silos and the barns, his pastures are scorched, and his animals bear within their panting bodies the anguish of drought. The fruit in his orchards falls to the ground, withered upon the stalk. Despairingly he identifies himself with cattle and pasture and corn, until he himself seems to crave moisture. If he were primitive man, or Pueblo Indian, he would offer victims before the altar of Baal, or set medicine men working for rain; he would find comfort upon the Arizona desert in the ritual of the rain dance.

But to him this relief is denied. Caught midway in man's evolution, between the solace of belief in imitative magic and the gropings of science for control over the weather, all he can do, this American farmer, is to wait with fevered patience, knowing that there is a law of averages, and that a crop failure now may be followed by filled ears and rich pastures next year.

But an evening comes when the farmer sees a restlessness in his animals. They feel rain, though the sky may be clear. The cattle seek shelter in the grove, and the hens cluster in a corner of the yard, clucking in agitation. The farmer searches the sky, but still it is without cloud. And then, as he turns to the house for supper, upon the horizon—what was it the Good Book said?—there he sees a cloud the size of a man's hand. Lying in bed that night, beside his wife, sleepless as they listen to the storm, he sees flashes of lightning play upon his corn fields, and hears rain slash the brittle stalks.

But now it is too late to swell the corn.

The farmer is helpless before wind and drought and flood. As he listens to the mockery of the rain upon his shrivelled corn, he remembers other droughts and other storms. The crop he had raised the first year of his marriage was as pretty a stand of corn as you could want. They had looked at it, on Labor Day, with contentment in their hearts. And then, before sundown, fifteen minutes of a hailstorm had laid their field low, and they had crept to bed that night humbled before a force greater than themselves.

But if this force is greater than man, it is also an exacting power. For it fires him with needs and urges that are beyond the growing of his food. Man belongs with the earth more closely than he remembers. He is a child of this earth, bound to it by an umbilical cord that he cannot sever. And so, as he goes about his labors, an unrest burns within him, beyond all reasoning.

Perhaps it is the sudden warmth in the air that has worked this magic. The common scenes of earth are illuminated by a glow of heroics, and the creator in him, which had been damped down by civilization, refuses to be subdued.

Out on the land the farmer walks among his animals. In a few days his sow will farrow. And then, but a short time after that, two of his mares should foal. Lambs gambol upon the meadow, and calves nozzle their mothers. As he walks among his stock, the farmer feels, in his creative instinct, the stirrings of content.

Inside the barn are sacks of seed. He cuts open the sacks, tenderly fingering the seeds within. At this moment he knows enchantment, for his eyes pass over reality and see in these multitudinous seeds the truth that lies behind fact, so that the microscopic kidney-shaped seeds of the clover, dull grey-black in color, dull red, pink, yellow, warm green, hold the magic of all seed for all time. He fingers them with wonder, stripped of the film of custom. They are smooth and cool and, dipping his hand deep into the sack, he knows that he holds within this hand an entire field of clover, and the scent of the blossom is strong in his nostrils and the hum of bees loud in his ears. At this moment, here in the barn among the sacks of seed, he becomes a god; and there is no bound to his power. Feeling this, an urge comes upon him to fling his arm in the gesture of sowing; for an urge does not die in man merely from disuse; it remains, to demand fulfilment. He fills his hand with the seed, and, stepping from the barn, throws out his right arm with a movement that he himself has never needed to use, but a movement that was instinct within the arms of his forefathers.

The farmer's wife works in the kitchen. But in her mind at this moment is the look in the farmer's face as he left the house for the yard. A wise woman knows when her man has gone away from her. She knows the look that comes to his eyes when he needs to create horses and cows, sheep and pigs, and a field of clover beyond the barn.

"The warmth in the air has got him again," she sighs. And she turns back to her own work. But between her and the kitchen stove she sees a man caught up into the fury of the creator who must have, at this moment, all earth to shape and all animals to mold. For there are times when the divinity in man possesses him, and he needs to sow seed over the face of the earth, whether it be upon the fields of his farm, or in a little patch of soil in a back yard. This divinity, that holds the power to transform one common man into a creator, finds little trouble in forming a boundless universe from a few feet of earth, enclosed by a fence. All man needs, to appease this god within him, is the chance to sow a seed within some earth.

Around him, else, the countryside will mock him. It will mock him in the burst of leaves on the trees, and the sprouting of shoots upon the earth. He will feel himself an outcast, playing no part in this rush of earth power. And he will not belong.

In his loneliness man needs to belong. He needs to be woven into the fabric of life. He cannot carry the solitude of exclusion; he craves common experience and shared delights. For only so can he feel sure of his own worth.

But there are deeper things he needs to share, and emotions he is scarcely aware of. These are the rhythms of the earth's year, which beat through his spirit with a pulse that he has forgotten to understand; for they have been overlaid by civilization, and lie beneath his consciousness. But they are not dead, for a few generations are little in the balance against the span of human existence, and æons of servitude to the earth weigh heavy. Vaguely he is made aware of these emotions, as the stores are filled in spring with brightly colored packets of seeds. It is no mere utilitarian reason that sends him out to his yard, with spade and hoe. Playing thus with earth and seeds, he is secure in his own worth, and feels no need to justify himself before mankind. In willingness he serves the pattern of life.

In the cities of the world, men sit at desks, working with figures. Within their dimmed spirits they, too, carry divinity, and the same unrest stirs their blood. They do not even know what irks and frets, so bounded are they by civilization. They cannot create animals and fields of clover. But life has made them in the image of God; it has given them the desire to create. They carry within these desk-bound bodies the fire of Prometheus. But it does not flame. It smolders, burning their spirits. And because they know that there is an unrest within them that they cannot still, they look with resentment at the small patch of window-framed sky. In some unformed awareness they feel themselves excluded, and they grow defiant, needing to justify themselves before men. This, here, is the birth of the will to power.

In the spring of the year, earth-force sent man out to sow seed upon the land. Now, as the earth is covered with the fruits of his toil, he is fired by the urge to harvest. This is a gregarious urge; for though he may sow his seed in solitude, yet he gathers his harvests riotously, in the company of his fellows. The land, that lay so quiet and still over the months of growth, resounds suddenly with the shouts and laughter of harvesters. Man obeys an ageless instinct which is as strong within him today, when airplanes cross the sky and grain is harvested by the combine, and armies battle against each other in tanks, as it was when he paid tribute in song and dance to the mythological gods of the earth.

Perhaps the urge of the harvest holds this power because it is unchanging. In a world of continual shift and destruction, it is an instinct that endures, beyond time and place. At this festival man can merge his separate identity into the triumph of all men, anywhere, since life began and the first crops and fruits were gathered. He can over-ride his own frontier and his own century, until he is cleansed of nationalism and moment, and becomes a human who has planted and toiled, and now harvests the fruits of his labors. He is accepted into the company of all tillers of the soil, from all time.

A chain of harvests encircles the world, heeding neither race nor creed. This chain leaps oceans and frontiers, slipping into enemy territory. For today there are two harvests; those upon the fields of the earth, and those within the longings of men. In a pause of the firing, in the fatigue of battle, the fighting man will see suddenly before his eyes the land of his own home; this land will be peopled with harvesters, and he, among them, will be hauling the last sheaves of wheat from the bottom field, or climbing the ladders against the apple trees in the orchard. And it may be that the field and the orchard lie upon that earth called America, or in that island called England; it may be that the apples ripen upon German trees, or the sheaves stand shocked in a field in Italy, bordered with grape vines. Within each fighting man stirs the urge of the harvest, and the need to rejoice. For man needs his festivals and his rituals, whether it be a corn husking in the Middle West, or the treading of the grapes in Provence, or a harvest festival in an English village church.

We hold these impulses in trust. We must guard them against the buffetings of a mechanized age. One day the armies that trample the fields of Europe will be withdrawn, and land will be dedicated again to growth. And when this happens, men will return to their own earth, scattered though it may be over the face of the globe. They will return to their fields and their farms, and we must see that the fields and farms of actuality are as fair as the fabric of their dreams. But there is something else that we must hand to these men. It is something very elusive and very ancient. It is older than history, and more rewarding than victory in battle. We must hand to them, tilled and cared for, the fields of fancy and the spirit, that they may renew the desire to sow their seed and obey the bidding of the seasons in the riot of harvesting.

For there are harvests of this earth as vital as corn and wheat, invisible though they may be to the human eye.

Insecure upon this globe, man needs to feel that there is something beyond alteration. The earth remaineth, sings the poet, and man alone passes. Man must believe this, even though science should disprove it. And fundamentally he is right, for though the soil which clothes this globe is a shifting substance, yet the spirit of the earth endures. Soil may trickle down the mountain sides, in a heavy rain, and be carried far from the place where first it disintegrated out of rock and decayed vegetable matter; but greater than the shift of soil is the unswerving pattern of earth behavior. The mountains may be the national bank of potential soil, and through the history of the earth lofty peaks repeatedly may have been raised above the level of the plains, and again worn away; but the common man cannot look upon life thus; he must see his earth as it is during his lifetime, and the lifetime of his father. There, to the west, the hills have stood, and always the sun has set behind them. The tales of his father, and the legends of his pioneer grandfathers, speak of those same hills, against the setting sun. He is not able to think of life as something that stretches back across the ages and will be flung forward to uncountable generations of his unborn children. He needs the sense of the moment, and the urgency of the visible present.

But even with this sense of the moment, the common man carries a feeling of service to humanity. Tell him that he works his soil in trust, and that the way he plows and sows now will determine the harvests of the future, and he will respond. Talk to him of sun and air and earth and water, and he will feel the magic with which he labors. For the man who works the earth labors with the tools of the gods, and his background is outside of time. What he does this season with his twenty-acre field will help to stamp the earth-pattern of future centuries; and all he sows and harvests is eternal. Let one farmer leave a field untended, he will have made a scar upon the earth long after the stone marking his grave has been softened with lichen. For the man who tills the earth is in service to the earth; and not merely to his own small portion of the globe, but to all earth. Boundaries are manmade, and of little significance. Soil-erosion knows no frontier, and obeys no landlord. It can scoop the life-giving topsoil from one man's land and deposit it upon the territory of a farmer a few hundred miles down the river, where the waters have overflowed their banks. And it can carry with it, upon its passage of disintegration, the land of a farmer who had striven to care for his soil. For as impartial winds can blow across the countryside weeds from an untended field, and a distant farmer will find an invasion of dock among his wheat, so may an undrained piece of land flood a neighbor's corn crop. "He can throw the water on me," says the Illinois farmer, watching with anxiety the careless planning of his neighbor's tiling. There is an inescapable brotherhood among the workers upon the land. For the earth remaineth only so long and so far as it is tended by all men upon all land.

A new element of excitement enters into this conception of an unstatic earth. The soil that lies upon the surface of our land becomes something to cherish, by reason of its mutability. We watch a drama that goes back to the beginning of time, and will never end. Upon the summits of the mountains the rocks crumble, gnawed by water and air, heat and cold. They merge with the decaying roots of the scrubby trees, and the remains of worms and all living things. They are soaked by rains and snows, to be rushed down the mountain-sides into the little streams. They are blown by the winds, to sift upon the fields in the valleys below. They shift, with the essence of all goodness in them; and they are called soil.

This cherished thing called soil is not imperishable or permanent. Neither does it lie thick upon the surface of our earth. It is a film only a few inches in depth, to be blown by the wind, wrenched away by floods, or knit together by the fibres of plants or the great roots of trees. For the tangle of plant life upon the soil is a close-woven mantle that binds this soil. In the Dust Bowl, where the plow upturned the mesh of prairie grass, that man in his greed might plunder all earth, nothing remained to check the shift of this loosened soil in the wind. But let nature have her way, and she will repair. Sid Edmonston will tell you of his farm: "Eight hundred of the prettiest acres I've got out there, and it was the toughest, orneriest piece of land during the time of the Dust Bowl. But nature and the native grass have brought it back."

We must carry with us an awareness of the perishability of this land, and learn to serve, rather than pillage, it. For nature has created her own rhythm, in which all forces are interconnected. And we have broken that rhythm. Our accelerated time sense wars with the time sense of the ages, and we have set a standard for production from the land that upsets the balance of nature. In our avarice we have removed forests and plowed up grasslands, confident that we were the masters. We did not know that the Nemesis of flood and soil erosion stood waiting. Over the South, gullies scar the land that once stood white with cotton. Out in Nebraska, the dust has blown from the earth, carrying all goodness with it. Upon the rangeland, the sheep and cattle clip the pastures clean, and the grass perishes. The people who work the land have been betrayed by their own misuse. They have been betrayed by the misuse of forebears, who in a new world, where land held no frontiers, felt no need to rebuild exhausted soil. These descendants of the pioneers walk their fields today with pinched faces and hopeless spirits. For the roots of their stability and plenty have withered beneath them. Their earth refuses to bear. Upon the same spot on this earth stands the cabin of their fathers; the same sky is above them; they are surrounded by the same wind and sun and rain. Everything remains except the one most vital element; the life-giving topsoil has been stripped from their land and lies, perhaps, out at sea beyond New Orleans, where it has been carried by the Father of Waters, in flood.

The people of a land reflect that land. And people with pinched faces and hopeless spirits have no excitement in freedom and security and the dream of democracy. They are poor white trash, lacking the will to fight.

A battle lies before us. It is not a war for new frontiers, for America stretches from ocean to ocean, and there are no new frontiers to seize. It is more in the nature of a crusade. It is the battle for the re-pioneering of the land, that this great stretch of earth, embracing all possible range of climate and soil, may grow sleek and fair, with neither barrenness nor flaw, and bear upon its varied surface a race of confident workers.

There are signs that this already is happening. Sid Edmonston tells us that his land in the Dust Bowl is "the prettiest country today." An airplane journey over the Southern States shows exciting patterns of contour plowing, saving the soil from the plunder of the rains. Tiny newly planted trees dot the gullied hillsides, that unborn generations may be spared the disaster of flood, and great man-made lakes and dams hold proof of our new mastery over the land.

But an even greater re-pioneering has begun. It is the re-pioneering of the spirit of man. We watch a swing to the earth. The city knows today that it holds less security than the farm, and, aware of their own invulnerability, the farmers themselves cease to feel inferior when they visit the cities. They are a community with pride, conscious of their independence. The call of the city holds little lure when the farmer works good earth, and money itself has less power in the threatened world of today. Man craves most the security of food.

In the foothills of the Blue Ridge Mountains a family farms a few acres of land. The farmer and his wife raise cows and steers, hogs and chickens. They grow corn and wheat, potatoes and vegetables. In the orchard behind the old stone house their trees bear fruit. The barn is filled with hay, the granary bulges with feed. Up in the meathouse hang hams. The cellar is stacked with canned fruit and vegetables. Go down the stone steps to the spring-house, and within the heavy door, in the cool stream, you will see butter and milk. Here is a land flowing with plenty, a world independent of the city. This farmer and his wife know that they need bow to no man, for they have, within the confines of their own earth, all food and drink. And because they till and plant their own earth, the spirit of restlessness has left them, and they have become members of a community, with a sense of service to that community.

Let the farmer call some land his own and he will love it and serve it; and loving it, he will stay with it. What remains of our earth is not the actual sifted atoms of soil, or the unchanged slope of a hill; it is something of far greater importance: it is the spirit of that soil. It is the immutable order of the earth, impressing its pattern upon man and giving him something to serve. For man needs order, and he is lonely if he is completely free. Seeing the order of the heavens, with the rising and setting of the punctual stars, and the rigid pattern of the earth, with the swing of the seasons; seeing, too, order behind the rhythm of dance and song, man understands that all life is order. Let him accept this, and within the design of this order he has freedom to laugh and sing. For he has lost the inhibiting spectre of fear. Germane abandon is possible only within this structure. The spans of a great bridge spring from order. The building of a cathedral is controlled by order. The birth of a child is possible only through order. So the farmer, who tills and plants his land, lives a life based on sanity; for he lives within the severest framework of order.

As we watch the farmer among his animals, or at plow, we know that we look upon a vicar of the eternal, a server of the gods who uses the elements as tools of his craft. We watch a man who holds our own being within his hands: for without him we should not be fed. But because his tools are the heavens and the earth, this man is without the arrogance he else might have.

For he knows he is servant to forces beyond human control. He is part of the whole of life, able to bewitch the common day with magic. He holds the secret of joy. Working with earth and rain, heat and seed and drought, he knows that "the pastures are clothed with flocks; the valleys also are covered over with corn; they shout for joy, they also sing."

S521.5.A2 L43 1971
Leighton / Give us this day,